LONDON MATHEMATICAL SOCIETY STUDENT TEXTS

43 Fourier analysis on finite groups and applications, AUDREY TERRAS
44 Classical invariant theory, PETER J. OLVER
45 Permutation groups, PETER J. CAMERON
47 Introductory lectures on rings and modules. JOHN A. BEACHY
48 Set theory, ANDRÁS HAJNAL & PETER HAMBURGER. Translated by ATTILA MATE
49 An introduction to K-theory for C*-algebras, M. RØRDAM, F. LARSEN & N. J. LAUSTSEN
50 A brief guide to algebraic number theory, H. P. F. SWINNERTON-DYER
51 Steps in commutative algebra: Second edition, R. Y. SHARP
52 Finite Markov chains and algorithmic applications, OLLE HÄGGSTRÖM
53 The prime number theorem, G. J. O. JAMESON
54 Topics in graph automorphisms and reconstruction, JOSEF LAURI & RAFFAELE SCAPELLATO
55 Elementary number theory, group theory and Ramanujan graphs, GIULIANA DAVIDOFF, PETER SARNAK & ALAIN VALETTE
56 Logic, induction and sets, THOMAS FORSTER
57 Introduction to Banach algebras, operators and harmonic analysis, GARTH DALES *et al*
58 Computational algebraic geometry, HAL SCHENCK
59 Frobenius algebras and 2-D topological quantum field theories, JOACHIM KOCK
60 Linear operators and linear systems, JONATHAN R. PARTINGTON
61 An introduction to noncommutative Noetherian rings: Second edition, K. R. GOODEARL & R. B. WARFIELD, JR
62 Topics from one-dimensional dynamics, KAREN M. BRUCKS & HENK BRUIN
63 Singular points of plane curves, C. T. C. WALL
64 A short course on Banach space theory, N. L. CAROTHERS
65 Elements of the representation theory of associative algebras I, IBRAHIM ASSEM, DANIEL SIMSON & ANDRZEJ SKOWROŃSKI
66 An introduction to sieve methods and their applications, ALINA CARMEN COJOCARU & M. RAM MURTY
67 Elliptic functions, J. V. ARMITAGE & W. F. EBERLEIN
68 Hyperbolic geometry from a local viewpoint, LINDA KEEN & NIKOLA LAKIC
69 Lectures on Kähler geometry, ANDREI MOROIANU
70 Dependence logic, JOUKU VÄÄNÄNEN
71 Elements of the representation theory of associative algebras II, DANIEL SIMSON & ANDRZEJ SKOWROŃSKI
72 Elements of the representation theory of associative algebras III, DANIEL SIMSON & ANDRZEJ SKOWROŃSKI
73 Groups, graphs and trees, JOHN MEIER
74 Representation theorems in Hardy spaces, JAVAD MASHREGHI
75 An introduction to the theory of graph spectra, DRAGOŠ CVETKOVIĆ, PETER ROWLINSON & SLOBODAN SIMIĆ
76 Number theory in the spirit of Liouville, KENNETH S. WILLIAMS
77 Lectures on profinite topics in group theory, BENJAMIN KLOPSCH, NIKOLAY NIKOLOV & CHRISTOPHER VOLL
78 Clifford algebras: An introduction, D. J. H. GARLING
79 Introduction to compact Riemann surfaces and dessins d'enfants, ERNESTO GIRONDO & GABINO GONZÁLEZ-DIEZ
80 The Riemann hypothesis for function fields, MACHIEL VAN FRANKENHUIJSEN
81 Number theory, Fourier analysis and geometric discrepancy, GIANCARLO TRAVAGLINI
82 Finite geometry and combinatorial applications, SIMEON BALL

London Mathematical Society Student Texts 83

The Geometry of Celestial Mechanics

HANSJÖRG GEIGES
University of Cologne

CAMBRIDGE
UNIVERSITY PRESS

University Printing House, Cambridge CB2 8BS, United Kingdom

Cambridge University Press is part of the University of Cambridge.

It furthers the University's mission by disseminating knowledge in the pursuit of education, learning and research at the highest international levels of excellence.

www.cambridge.org
Information on this title: www.cambridge.org/9781107125407

First published 2016

A catalogue record for this publication is available from the British Library

Library of Congress Cataloguing in Publication data
Names: Geiges, Hansjörg, 1966–
Title: The geometry of celestial mechanics / Hansjörg Geiges, University of Cologne.
Description: Cambridge : Cambridge University Press, 2016. | Series: London Mathematical Society student texts ; 83 | Includes bibliographical references and index.
Identifiers: LCCN 2015038450 | ISBN 9781107125407 (Hardback : alk. paper) | ISBN 9781107564800 (Paperback : alk. paper)
Subjects: LCSH: Celestial mechanics. | Kepler's laws.
Classification: LCC QB355.3 .G45 2016 | DDC 521–dc23 LC record available at http://lccn.loc.gov/2015038450

ISBN 978-1-107-12540-7 Hardback

Und in der Tat, die wichtigsten geistigen Vorkehrungen der Menschheit dienen der Erhaltung eines beständigen Gemütszustands, und alle Gefühle, alle Leidenschaften der Welt sind ein Nichts gegenüber der ungeheuren, aber völlig unbewußten Anstrengung, welche die Menschheit macht, um sich ihre gehobene Gemütsruhe zu bewahren! Es lohnt sich scheinbar kaum, davon zu reden, so klaglos wirkt es. Aber wenn man näher hinsieht, ist es doch ein äußerst künstlicher Bewußtseinszustand, der dem Menschen den aufrechten Gang zwischen kreisenden Gestirnen verleiht und ihm erlaubt, inmitten der fast unendlichen Unbekanntheit der Welt würdevoll die Hand zwischen den zweiten und dritten Rockknopf zu stecken.

Robert Musil, *Der Mann ohne Eigenschaften*

Contents

Preface

> DER INQUISITOR Und da richten diese Würmer von Mathematikern ihre Rohre auf den Himmel [...] Ist es nicht gleichgültig, wie diese Kugeln sich drehen?
>
> Bertolt Brecht, *Leben des Galilei*

Celestial mechanics has attracted the interest of some of the greatest mathematical minds in history, from the ancient Greeks to the present day. Isaac Newton's deduction of the universal law of gravitation (Newton, 1687) triggered enormous advances in mathematical astronomy, spearheaded by the mathematical giant Leonhard Euler (1707–1783). Other mathematicians who drove the development of celestial mechanics in the first half of the eighteenth century were Alexis Claude Clairaut (1713–1765) and Jean le Rond d'Alembert (1717–1783), see (Linton, 2004). In those days, the demarcation lines separating mathematics and physics from each other and from intellectual life in general had not yet been drawn. Indeed, d'Alembert may be more famous as the co-editor with Denis Diderot of the *Encyclopédie*. During the Enlightenment, celestial mechanics was a subject discussed in the salons by writers, philosophers and intellectuals like Voltaire (1694–1778) and Émilie du Châtelet (1706–1749).

The history of celestial mechanics continues with Joseph-Louis Lagrange (1736–1813), Pierre-Simon de Laplace (1749–1827) and William Rowan Hamilton (1805–1865), to name but three mathematicians whose contributions will be discussed at length in this text. Henri Poincaré (1854–1912), perhaps the last universal mathematician, initiated the modern study of the three-body problem, together with large parts of the theory of dynamical systems and what is now known as symplectic geometry (Barrow-Green, 1997; Charpentier *et al.*, 2010; McDuff and Salamon, 1998).

Yet this time-honoured subject seems to have all but vanished from the mathematical curricula of our universities. This is reflected in the available textbooks, which are either getting a bit long in the tooth, or are addressed to a fairly advanced and specialised audience. The *Lectures on Celestial Mechanics* by Siegel and Moser (1971), a classic in their own right, deal with Sundman's work on the three-body problem in the wake of Poincaré's, and with questions about periodic solutions and stability, all at a rather mature level. Celestial mechanics as a key motivation for the study of dynamical systems is served well by (Moser and Zehnder, 2005) and (Meyer *et al.*, 2009).

My personal interest in celestial mechanics stems from reading the paper (Albers *et al.*, 2012), where the three-body problem is approached with methods from contact topology, my core area of expertise, see (Geiges, 2008). I should say 'attempting to read', for I quickly realised that I was ignorant of some of the most basic terminology in celestial mechanics.

In order to remedy this deplorable state of affairs – and to confute the inquisitor – I decided to teach a course on celestial mechanics, with (Pollard, 1966), (Danby, 1992) and (Ortega and Ureña, 2010) as my excellent guides. The latter textbook can be recommended even to readers whose grasp of Spanish is as rudimentary as mine.

However, none of these texts takes the geometric view that I wished to emphasise, so I included material from sources such as (Milnor, 1983) and (Hall and Josić, 2000), expanded and adapted to the needs of an introductory course. The present text rather faithfully reflects the course I taught at the University of Cologne in 2012/13, where the audience of some seventy ranged from second-year mathematics or physics undergraduates all the way to Ph.D. students. For a follow-up seminar in 2014/15 and this write-up I added more geometric material, notably on the curvature of planar curves and projective geometry, inspired by (Coolidge, 1920), and I removed a couple of sections on generating functions and Hamilton–Jacobi theory, which I felt were less in the spirit of this elementary geometry course in disguise.

The result, I hope, is a text that can be read profitably by undergraduates in their penultimate or final year, while not being too pedestrian for more advanced students. I believe that, for students not intending to specialise in geometry, learning elementary differential geometry and topology by seeing it 'in action', that is, applied to questions in celestial mechanics, may be a more satisfying experience than some traditional courses that concentrate on the development of machinery and often stop before the student can really appreciate its utility – needless to say, students who plan to continue with further courses in geometry may likewise enjoy that experience. Celestial mechanics is a field where many strands of pure and applied mathematics come together, and for this reason alone it deserves a more prominent place in the curriculum.

I have included over a hundred exercises, often with comments that explain their relevance, making the text suitable for self-study. It should be possible to cover most of this book in a one-semester course of 14 weeks. For shorter courses one could omit the proof of planarity in Lagrange's theorem (Theorem 7.1) and make a selective choice of the material in Chapters 8 to 10.

The contents of this book

A large portion of this text is concerned with the simplest question in celestial mechanics, the Kepler problem, which studies the motion of a single body around a fixed centre under Newtonian attraction. One of my aims is to display the rich geometry of this problem. In particular, several proofs of Kepler's first law about the shape of the orbit will be given, based on geometric concepts such as curvature of planar curves or conformal (i.e. angle-preserving) transformations of the plane.

Chapter 1 introduces the central force problem, where the force law need not be Newtonian. Even in this more general setting one finds two preserved quantities of the motion: the angular momentum and, if the force field derives from a potential, the energy. The preservation of the angular momentum can be rephrased as Kepler's second law about areas.

Kepler's first law about the shape of the orbit, now assuming Newtonian attraction, is proved (following Laplace) in Chapter 3: the orbit is a conic section, with one focus in the force centre. Chapter 2 provides the background on conic sections, to which the reader may refer as needed.

Of course, knowing the shape of the orbit is only half the answer, in particular if you are trying to locate a celestial object in the sky. One would really like to have an explicit time parametrisation of the orbit. This surprisingly difficult question is the theme of Chapter 4. In the elliptic and hyperbolic case it leads to a transcendental equation named after Kepler; I present a geometric solution of this equation, due to Newton, involving a famous planar curve, the cycloid. In the parabolic case it leads to a cubic equation, and I reveal the geometry behind the algebraic solution of such equations.

Passing from one to two bodies moving under mutual attraction, we shall see in the brief Chapter 5 that this question reduces quite easily to the Kepler problem.

Chapter 6 investigates the central question of celestial mechanics, the n-body problem: How do n point masses move in $\mathbb{R}^3$ under mutual Newtonian attraction? We find some preserved quantities of this problem that allow us to make certain statements about the long-time behaviour of n-body systems, although we remain far from finding concrete solutions. In a section on central configurations I exhibit explicit solutions under additional geometric assumptions.

Chapter 7 deals with the special case $n = 3$. The centre-piece of that chapter is Lagrange's beautiful theorem on homographic (i.e. self-similar) solutions of the three-body problem. I also discuss the restricted three-body problem, where one of the three masses is negligibly small compared with the others.

In Chapter 8 we return to the Kepler problem, but from a more geometric point of view. This is really the geometric heart of the present text, where several types of geometric transformations (inversion, stereographic projection, polar reciprocation), spaces (hyperbolic space, projective plane) and differential geometric concepts (geodesics, curvature, conformal maps) are introduced. These techniques are used not only to give alternative proofs of Kepler's first law, but chiefly to give a unified view of all Kepler solutions, including the collision orbits (theorems of Moser, Osipov, and Belbruno).

Chapter 9 prepares the reader for the modern literature on the n-body problem by introducing the Hamiltonian formalism, starting from variational principles. In Chapter 10, the Hamiltonian formalism is applied to the Kepler problem. We determine the topology of the three-dimensional energy hypersurfaces in this problem, and I present a number of equivalent topological descriptions of these 3-manifolds. In particular, I use the quaternions to identify the special orthogonal group SO(3) as projective 3-space. Energy hypersurfaces with this topology also arise in the restricted three-body problem.

All chapters but one end with extensive historical notes and references.

Notational conventions

Vector quantities will be denoted in bold face; the euclidean length of a vector quantity is usually denoted by the corresponding symbol in italics. For example, $\mathbf{r}$ denotes the position vector of a particle in $\mathbb{R}^3$, and $r := |\mathbf{r}|$. The norm $|\,.\,|$ will always be the euclidean one. The standard (euclidean) inner product on $\mathbb{R}^3$ will be denoted by $\langle\,.\,,\,.\,\rangle$.

Time derivatives will be written with dots in the Newtonian fashion. For instance, if $t \mapsto \mathbf{r}(t)$ denotes the motion of a particle, its velocity $\mathbf{v}$ and acceleration $\mathbf{a}$ are given by

$$\mathbf{v} := \dot{\mathbf{r}} := \frac{\mathrm{d}\mathbf{r}}{\mathrm{d}t}, \quad \mathbf{a} := \ddot{\mathbf{r}} := \frac{\mathrm{d}^2\mathbf{r}}{\mathrm{d}t^2}.$$

The length $v := |\mathbf{v}|$ of the velocity vector is called the speed.

The natural numbers $\mathbb{N}$ are the positive integers; if 0 is to be included, I write $\mathbb{N}_0$. The rational, real and complex numbers are denoted by $\mathbb{Q}$, $\mathbb{R}$ and $\mathbb{C}$, respectively. The positive reals are denoted by $\mathbb{R}^+$; the negative reals, by $\mathbb{R}^-$. We set $\mathbb{R}_0^+ := \mathbb{R}^+ \cup \{0\}$ and $\mathbb{R}^\times := \mathbb{R} \setminus \{0\}$. The notation $\mathbb{H}$ stands for hyperbolic space or Hamilton's quaternions, depending on the context. I use the standard notation C^k, $k \in \mathbb{N}$, for k times continuously differentiable functions or maps. By C^0 I simply mean continuous. Functions or maps of class C^∞ are also referred to as **smooth**.

Physical background

No prior knowledge of physics will be assumed apart from the following two Newtonian laws.

The second Newtonian law of motion: The acceleration $\mathbf{a}$ experienced by a body of mass m under the influence of a force $\mathbf{F}$ is given by

$$\mathbf{F} = m\mathbf{a}.$$

The universal law of gravitation: The force exerted by a body of mass m_2 at the point $\mathbf{r}_2 \in \mathbb{R}^3$ on a body of mass m_1 at the point $\mathbf{r}_1 \in \mathbb{R}^3$ equals

$$\mathbf{F} = \frac{Gm_1m_2}{r^2} \cdot \frac{\mathbf{r}}{r},$$

where $\mathbf{r} := \mathbf{r}_2 - \mathbf{r}_1$, and

$$G \approx 6.673 \cdot 10^{-11} \frac{\mathrm{m}^3}{\mathrm{kg\ s}^2}$$

is the universal gravitational constant.

Mathematical background

I have tried to keep the mathematical prerequisites to a minimum, but the level of sophistication certainly increases as this text proceeds. A great number of the students taking my class at the University of Cologne were physics undergraduates in the second year of their studies. In their first year, they had followed my course on analysis and linear algebra, where they had seen, amongst other things, basic topological concepts, the notion of local and global diffeomorphisms, the inverse and the implicit function theorems, the classical matrix groups, elementary ordinary differential equations (the Picard–Lindelöf theorem on local existence and uniqueness, linear systems with constant coefficients), submanifolds, the transformation formula for higher-dimensional integrals, the integral theorems of Gauß and Stokes, and differential forms. In this text, submanifolds make a brief appearance in Chapter 7 and in the exercises to Chapter 8; the concept is essential for Section 9.2 and Chapter 10. Differential forms are used only in Section 9.2. Homeomorphisms (i.e. bijective maps that are continuous in either direction) and diffeomorphisms between submanifolds make a brief appearance in Section 8.3, and they become central only in Chapter 10. In that last chapter I also assume a certain familiarity with basic notions in point-set topology (Hausdorff property, compactness); the relevant material can be found in (Jänich, 2005) or (McCleary, 2006). In the context of an alternative proof of Kepler's first law, holomorphic maps appear in a couple of isolated places in the exercises to Chapter 8 and in Section 9.1.

As regards differential equations, throughout I use the following geometric interpretation. Let $\Omega \subset \mathbb{R}^d$ be an open subset and $\mathbf{X}$ a **vector field** on Ω, i.e. a function $\mathbf{X}\colon \Omega \to \mathbb{R}^d$. This gives rise to a first-order differential equation

$$\dot{\mathbf{x}} = \mathbf{X}(\mathbf{x}).$$

Solutions of this differential equations are C^1-maps $\mathbf{x}\colon I \to \Omega$, defined on some interval $I \subset \mathbb{R}$, that satisfy this equation; that is,

$$\dot{\mathbf{x}}(t) = \mathbf{X}(\mathbf{x}(t)) \quad \text{for all } t \in I.$$

In geometric terms this means that $\mathbf{x}$ is an **integral curve** or **flow line** of $\mathbf{X}$, i.e. a curve whose velocity vector $\dot{\mathbf{x}}(t)$ at the point $\mathbf{x}(t)$ coincides with the vector $\mathbf{X}(\mathbf{x}(t))$ defined by the vector field $\mathbf{X}$ at that point.

The Picard–Lindelöf existence and uniqueness theorem (known to French readers as the Cauchy–Lipschitz theorem) says that if $\mathbf{X}$ is locally Lipschitz continuous, then for any $\mathbf{x}_0 \in \Omega$ the initial value problem

$$\dot{\mathbf{x}} = \mathbf{X}(\mathbf{x}), \quad \mathbf{x}(0) = \mathbf{x}_0$$

has a solution defined on some small time interval $(-\delta, \delta)$, and two such solutions coincide on the time interval around 0 where both are defined. In all cases studied in this text, the vector field will actually be C^1 (or even smooth), so that local Lipschitz continuity is guaranteed by the mean value theorem.

Excellent texts on differential equations emphasising the geometric viewpoint are (Arnol'd, 1973) and (Bröcker, 1992). I can also recommend (Givental, 2001) and (Robinson, 1999). An eminently readable proof of the Picard–Lindelöf theorem is given in Appendix A of (Borrelli and Coleman, 2004).

Permissions

I am grateful for permission to quote from the following works.

- Th. Bernhard, *Korrektur.* © Suhrkamp Verlag, Frankfurt am Main, 1975. All rights reserved by and controlled through Suhrkamp Verlag, Berlin.
- K. Bonfiglioli, *Something Nasty in the Woodshed.* © the Estate of Kyril Bonfiglioli, 1976. Used by kind permission of Mrs Margaret Bonfiglioli.
- B. Brecht, *Leben des Galilei.* © Bertolt-Brecht-Erben/Suhrkamp Verlag.

I thank Clare Dennison of Cambridge University Press for procuring these permissions.

Acknowledgements

I thank Roger Astley of Cambridge University Press for his enthusiasm when I approached him with my incipient ideas for this book, and the editors of the LMS Student Texts for accepting it into their series. I also thank my copy-editor Steven Holt for his excellent work.

The participants of both my lecture course and the student seminar on a manuscript version of this book forced me to clarify my thoughts. Sebastian Durst, who organised the seminar with me, found a number of misprints and inaccuracies. Marc Kegel and Markus Kunze were likewise eagle-eyed proofreaders, and they made valuable suggestions for improving the exposition. Alain Chenciner and Jacques Féjoz were especially helpful with references. Isabelle Charton (2013) wrote a Bachelor thesis on (Coolidge, 1920) under my supervision, translating that paper into the language of projective geometry. Although I follow a different route via polar reciprocation in Section 8.6, this part would not have been written were it not for the discussions with her. Max Dörner contributed Figure 4.2; Otto van Koert, Figure 7.4. Conversations with Peter Albers, Alain Chenciner, Jacques Féjoz, Pablo Iglesias, Bernd Kawohl, Otto van Koert, Markus Kunze, Janko Latschev, Rafael Ortega, Patrick Popescu-Pampu, Thomas Rot, Felix Schlenk, Karl Friedrich Siburg, Guido Sweers and Kai Zehmisch have left distinct traces in this book. I thank them all for making this a better text.

Cologne, November 2015 Hansjörg Geiges

1

The central force problem

Moment. Das Moment des Eindrucks, den ein Mann auf das gemeine Volk macht, ist ein Produkt aus dem Wert des Rocks in den Titel.

Georg Christoph Lichtenberg

We start by dealing with an idealised problem, where a body of mass m – more precisely, a dimensionless particle – is moving in euclidean 3-space $\mathbb{R}^3$ subject to an attractive force directed towards a fixed centre, which for convenience we place at the origin $\mathbf{0} \in \mathbb{R}^3$. To begin with, we allow more general forces than the one described by Newton's law of gravitation; the force may even be repelling in some regions of space.

The central force problem Find solutions of the differential equation[1]

$$m\ddot{\mathbf{r}} = -mf(\mathbf{r}) \cdot \frac{\mathbf{r}}{r}, \qquad \text{(CFP)}$$

where $f\colon\ \mathbb{R}^3 \setminus \{\mathbf{0}\} \to \mathbb{R}$ is a given continuous function.

By a solution of (CFP) we mean a C^2-map $\mathbf{r}\colon\ I \to \mathbb{R}^3 \setminus \{\mathbf{0}\}$, defined on some interval $I \subset \mathbb{R}$, that satisfies the differential equation (CFP).

The equation (CFP) expresses the requirement that the force act along the line joining the body and the centre $\mathbf{0}$, and that it be proportional to the mass m, so that in fact m is irrelevant to the solution of (CFP). The sign in (CFP) has been chosen in such a way that the force is attracting at points $\mathbf{r} \in \mathbb{R}^3 \setminus \{\mathbf{0}\}$ where $f(\mathbf{r}) > 0$, which is the case we usually consider, and repelling where $f(\mathbf{r}) < 0$.

As is standard in the theory of differential equations, the second-order differential equation (CFP) can be rewritten as a system of first-order equations

[1] All equations with individual labels are listed in the Index under 'equations'.

by introducing the velocity $\mathbf{v} := \dot{\mathbf{r}}$ as an additional variable:

$$\left.\begin{array}{rcl} \dot{\mathbf{r}} & = & \mathbf{v} \\ \dot{\mathbf{v}} & = & -f(\mathbf{r}) \cdot \dfrac{\mathbf{r}}{r}. \end{array}\right\} \qquad \text{(CFP}'\text{)}$$

Any C^2-solution $t \mapsto \mathbf{r}(t)$ of (CFP) gives rise to a C^1-solution $t \mapsto (\mathbf{r}(t), \dot{\mathbf{r}}(t))$ of (CFP′). Conversely, the first component of a C^1-solution $t \mapsto (\mathbf{r}(t), \mathbf{v}(t))$ of (CFP′) is a solution of (CFP); notice that the equations (CFP′) imply that this first component is actually of class C^2.

The advantage of the formulation (CFP′) is that we can think of solutions as integral curves of the vector field $\mathbf{X}$ on $(\mathbb{R}^3 \setminus \{\mathbf{0}\}) \times \mathbb{R}^3$ defined by

$$\mathbf{X}(\mathbf{r}, \mathbf{v}) = (\mathbf{v}, -f(\mathbf{r}) \cdot \mathbf{r}/r).$$

This point of view will become relevant in Sections 3.1 and 6.2, for instance.

A first step towards understanding a system of differential equations is to ask whether there are any preserved quantities. In the physical context considered here, these will be referred to as **constants of motion**. By this we mean functions of $\mathbf{r}$ and $\mathbf{v}$ that are constant along integral curves of (CFP′). In this chapter we shall meet two such constants of motion: the angular momentum and, in the centrally symmetric case, the total energy.

1.1 Angular momentum and Kepler's second law

Definition 1.1 Let $\mathbf{r}\colon I \to \mathbb{R}^3$ be a C^1-map. The **angular momentum** of $\mathbf{r}$ about $\mathbf{0} \in \mathbb{R}^3$ is

$$\boxed{\mathbf{c}(t) := \mathbf{r}(t) \times \dot{\mathbf{r}}(t),}$$

where $\times$ denotes the usual cross product in $\mathbb{R}^3$.

Remark 1.2 The physical angular momentum of a body of mass m moving along a trajectory described by the map $\mathbf{r}$ is $m\mathbf{r} \times \dot{\mathbf{r}}$, so – strictly speaking – our angular momentum is the angular momentum per unit mass.

Proposition 1.3 *In the central force problem, the angular momentum is a constant of motion. For $\mathbf{c} \neq \mathbf{0}$, the motion takes place in the plane through $\mathbf{0}$ orthogonal to $\mathbf{c}$. For $\mathbf{c} = \mathbf{0}$, the motion is along a straight line through $\mathbf{0}$.*

Proof With (CFP) we find

$$\dot{\mathbf{c}} = \dot{\mathbf{r}} \times \dot{\mathbf{r}} + \mathbf{r} \times \ddot{\mathbf{r}} = \mathbf{0}.$$

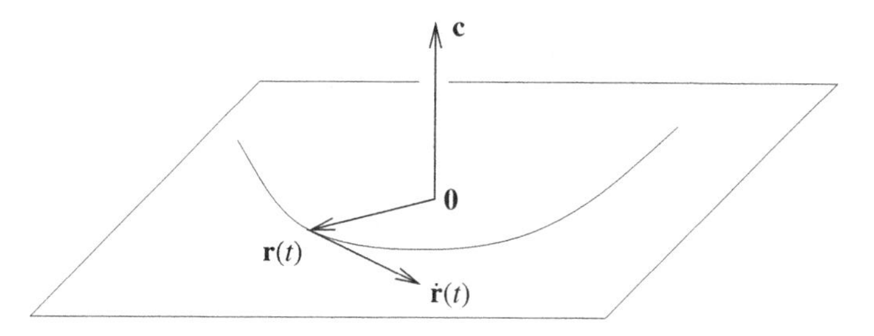

Figure 1.1 The angular momentum $\mathbf{c}$.

This proves the first statement. Moreover, we have $\langle \mathbf{c}, \mathbf{r} \rangle = 0$ by the definition of $\mathbf{c}$. This settles the case $\mathbf{c} \neq \mathbf{0}$, see Figure 1.1.

In order to deal with the case $\mathbf{c} = \mathbf{0}$, we first derive a general identity in vector analysis. Consider a C^1-map $\mathbf{u}\colon I \to \mathbb{R}^3 \setminus \{\mathbf{0}\}$. Then

$$\dot{u} = \frac{\mathrm{d}}{\mathrm{d}t}\sqrt{\langle \mathbf{u}, \mathbf{u} \rangle} = \frac{\langle \mathbf{u}, \dot{\mathbf{u}} \rangle}{\sqrt{\langle \mathbf{u}, \mathbf{u} \rangle}} = \frac{\langle \mathbf{u}, \dot{\mathbf{u}} \rangle}{u},$$

hence

$$u\dot{u} = \langle \mathbf{u}, \dot{\mathbf{u}} \rangle.$$

Then

$$\frac{\mathrm{d}}{\mathrm{d}t}\left(\frac{\mathbf{u}}{u}\right) = \frac{\dot{\mathbf{u}}u - \mathbf{u}\dot{u}}{u^2} = \frac{\dot{\mathbf{u}}\langle \mathbf{u}, \mathbf{u} \rangle - \mathbf{u}\langle \mathbf{u}, \dot{\mathbf{u}} \rangle}{u^3}.$$

With the **Graßmann identity**

$$(\mathbf{a} \times \mathbf{b}) \times \mathbf{c} = \langle \mathbf{a}, \mathbf{c} \rangle\, \mathbf{b} - \langle \mathbf{b}, \mathbf{c} \rangle\, \mathbf{a} \tag{1.1}$$

for vectors $\mathbf{a}, \mathbf{b}, \mathbf{c} \in \mathbb{R}^3$ (see Exercise 1.2), we obtain

$$\frac{\mathrm{d}}{\mathrm{d}t}\left(\frac{\mathbf{u}}{u}\right) = \frac{(\mathbf{u} \times \dot{\mathbf{u}}) \times \mathbf{u}}{u^3}.$$

Specialising to $\mathbf{u} = \mathbf{r}$, we get

$$\boxed{\frac{\mathrm{d}}{\mathrm{d}t}\left(\frac{\mathbf{r}}{r}\right) = \frac{\mathbf{c} \times \mathbf{r}}{r^3}.} \tag{1.2}$$

Hence, for $\mathbf{c} = \mathbf{0}$ the vector $\mathbf{r}/r$ is constant, and $\mathbf{r}(t) = r(t) \cdot \mathbf{r}(t)/r(t)$ is always a positive multiple of this constant vector. □

Given a solution to (CFP), we may choose our coordinate system in such a way that the motion takes place in the xy-plane. Therefore, for the remainder of this section, we restrict our attention to planar curves

$$\alpha\colon I \to \mathbb{R}^2 \setminus \{\mathbf{0}\}.$$

We write

$$\alpha(t) = (\alpha_1(t), \alpha_2(t)) = r(t) \cdot (\cos\theta(t), \sin\theta(t))$$

with $r(t) > 0$.

Remark 1.4 The transformation from polar to cartesian coordinates is described by the smooth map

$$\begin{array}{rccc} p\colon & \mathbb{R}^+ \times \mathbb{R} & \longrightarrow & \mathbb{R}^2 \setminus \{\mathbf{0}\} \\ & (r, \theta) & \longmapsto & (r\cos\theta, r\sin\theta). \end{array}$$

The Jacobian determinant of this map is

$$\det J_{p,(r,\theta)} = \begin{vmatrix} \cos\theta & -r\sin\theta \\ \sin\theta & r\cos\theta \end{vmatrix} = r \neq 0,$$

so p is a local diffeomorphism by the inverse function theorem. Moreover, p is a **covering map**, which means the following. For any point in $\mathbb{R}^2 \setminus \{\mathbf{0}\}$ one can find an open, path-connected neighbourhood U whose preimage $p^{-1}(U)$ is a non-empty disjoint union of sets U_λ, $\lambda \in \Lambda$, such that $p|_{U_\lambda}\colon U_\lambda \to U$ is a homeomorphism for each λ in the relevant index set Λ. (You are asked to verify this property in Exercise 1.3.)

The covering space property guarantees that any continuous curve $\alpha\colon I \to \mathbb{R}^2 \setminus \{\mathbf{0}\}$ can be lifted to a continuous curve $\tilde{\alpha}\colon I \to \mathbb{R}^+ \times \mathbb{R}$ with $p \circ \tilde{\alpha} = \alpha$, and this lift is uniquely determined by the choice of $\tilde{\alpha}(t_0) \in p^{-1}(\alpha(t_0))$ for some $t_0 \in I$. The local diffeomorphism property implies that if α was of class C^k, so will be $\tilde{\alpha}$.

The upshot is that a planar C^k-curve α can be written in polar coordinates with C^k-functions r and θ. Of course, r is uniquely determined by $r = |\alpha|$; the function θ is uniquely determined up to adding an integer multiple of 2π by an appropriate choice $\theta(t_0)$ at some $t_0 \in I$, and the requirement that θ at least be continuous.

In Exercise 1.4 you are asked to arrive at the same conclusion by an argument that does not involve any topological reasoning, but only the Picard–Lindelöf theorem.

Proposition 1.5 *Let*

$$\alpha = r(\cos\theta, \sin\theta)\colon [t_0, t_1] \longrightarrow \mathbb{R}^2 \setminus \{\mathbf{0}\}$$

be a C^1-curve with **angular velocity** $\dot{\theta} > 0$ *on* $[t_0, t_1]$ *and* $\theta(t_1) - \theta(t_0) < 2\pi$. *Then the area of*

$$D := \{s\alpha(t)\colon t \in [t_0, t_1],\ s \in [0, 1]\}$$

(see Figure 1.2) is given by

$$\text{area}(D) = \frac{1}{2}\int_{t_0}^{t_1} r^2(t)\,\dot{\theta}(t)\,\mathrm{d}t.$$

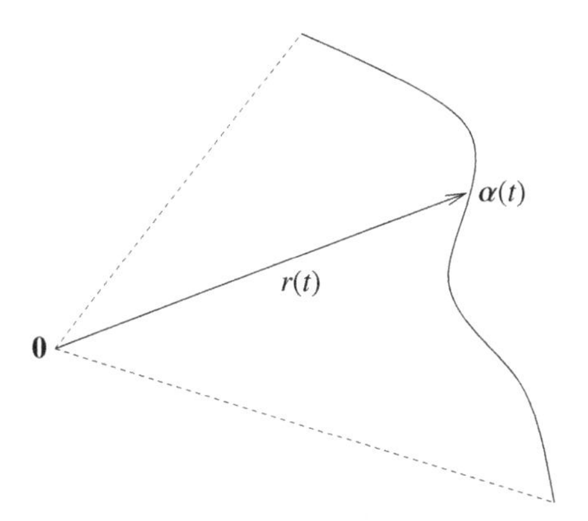

Figure 1.2 The area swept out by the position vector.

Proof Because of $\dot{\theta} > 0$, the inverse function theorem allows us to regard t and hence r as a function of θ. The area element $\mathrm{d}A$ is given in polar coordinates by $\mathrm{d}A = r\,\mathrm{d}r\,\mathrm{d}\theta$; this follows from the transformation formula:

$$\mathrm{d}A = \mathrm{d}x\,\mathrm{d}y = \begin{vmatrix} \partial x/\partial r & \partial x/\partial\theta \\ \partial y/\partial r & \partial y/\partial\theta \end{vmatrix} \mathrm{d}r\,\mathrm{d}\theta = r\,\mathrm{d}r\,\mathrm{d}\theta,$$

cf. Remark 1.4. Write $\theta_i = \theta(t_i)$, $i = 0, 1$. Then

$$\text{area}(D) = \int_{\theta_0}^{\theta_1}\int_0^{r(\theta)} \rho\,\mathrm{d}\rho\,\mathrm{d}\theta = \frac{1}{2}\int_{\theta_0}^{\theta_1} r^2(\theta)\,\mathrm{d}\theta.$$

The transformation rule (with $r(t) := r(\theta(t))$ and $\mathrm{d}\theta = \dot{\theta}\,\mathrm{d}t$) yields the claimed area formula. □

Theorem 1.6 (Kepler's second law) *The radial vector describing a solution of the central force problem sweeps out equal areas in equal intervals of time.*

Proof Choose coordinates such that the motion takes place in the xy-plane and $\mathbf{c}$ points in the positive z-direction. We may then write

$$\mathbf{r}(t) = r(t)\cdot(\cos\theta(t), \sin\theta(t), 0).$$

Hence

$$\dot{\mathbf{r}}(t) = (\dot{r}\cos\theta - r\dot{\theta}\sin\theta, \dot{r}\sin\theta + r\dot{\theta}\cos\theta, 0).$$

By Proposition 1.3, the angular momentum

$$\mathbf{c} = \mathbf{r} \times \dot{\mathbf{r}} = (0, 0, r^2\dot{\theta})$$

is constant. With our assumption on the orientation of $\mathbf{c}$ this implies $r^2\dot{\theta} = c$. (In particular, we have $\dot{\theta} \geq 0$, i.e. the motion is along a line or counter-clockwise in the xy-plane.) Thus, according to Proposition 1.5, the area swept out during the time interval $[t_0, t_1]$ equals $c(t_1 - t_0)/2$. □

Remark 1.7 The converse to this theorem is also true: if a planar motion in a force field satisfies Kepler's second law with respect to the centre $\mathbf{0}$, the force field is central. This can be seen as follows. If Kepler's second law holds, then $r^2\dot{\theta}$ is constant. By the computation in the proof above, this means that $\mathbf{c}$ is constant. Hence $\mathbf{0} = \dot{\mathbf{c}} = \mathbf{r} \times \ddot{\mathbf{r}}$, which means that $\ddot{\mathbf{r}}$ is parallel to $\mathbf{r}$.

1.2 Conservation of energy

The motion of a particle in an open subset $\Omega \subset \mathbb{R}^3$ under the influence of a force field $\mathbf{F}\colon \Omega \to \mathbb{R}^3$ is described by the Newtonian differential equation

$$m\ddot{\mathbf{r}} = \mathbf{F}. \tag{N}$$

The force field $\mathbf{F}$ is called **conservative** if it has a **potential** $V\colon \Omega \to \mathbb{R}$, i.e. a C^1-function such that

$$\mathbf{F} = -\operatorname{grad} V.$$

Proposition 1.8 *For motions in a conservative force field, i.e. solutions of the Newtonian differential equation*

$$m\ddot{\mathbf{r}} = -\operatorname{grad} V(\mathbf{r}), \tag{N$_c$}$$

the **total energy**

$$E(t) := \frac{1}{2}mv^2(t) + V(\mathbf{r}(t))$$

is constant.

Proof With $v^2 = \langle \dot{\mathbf{r}}, \dot{\mathbf{r}} \rangle$ we see that

$$\frac{\mathrm{d}E}{\mathrm{d}t} = m\langle \ddot{\mathbf{r}}, \dot{\mathbf{r}} \rangle + \langle \operatorname{grad} V(\mathbf{r}), \dot{\mathbf{r}} \rangle = \langle m\ddot{\mathbf{r}} + \operatorname{grad} V(\mathbf{r}), \dot{\mathbf{r}} \rangle = 0. \qquad \square$$

Remark 1.9 The **kinetic energy** $mv^2/2$ is the work required to accelerate a body of mass m from rest to velocity $\mathbf{v}$. Indeed, if we accelerate the body along a path γ during the time interval $[0, T]$ from rest to the final velocity $\mathbf{v}(T)$, the work done is

$$\int_\gamma \langle \mathbf{F}, \mathrm{d}\mathbf{s} \rangle = \int_0^T m \left\langle \mathbf{a}, \frac{\mathrm{d}\mathbf{s}}{\mathrm{d}t} \right\rangle \mathrm{d}t = \int_0^T m \langle \dot{\mathbf{v}}, \mathbf{v} \rangle \,\mathrm{d}t = \frac{1}{2} m |\mathbf{v}(T)|^2.$$

Example 1.10 The force field describing the *centrally symmetric* central force problem is conservative. In order to see this, notice that in the centrally symmetric case the function f in (CFP) depends only on r rather than $\mathbf{r}$, i.e. we have

$$\mathbf{F} = -mf(r) \cdot \frac{\mathbf{r}}{r}.$$

I claim that the potential of this force field is given by the centrally symmetric function

$$V(r) := m \int_{r_0}^{r} f(\rho) \,\mathrm{d}\rho.$$

Indeed, we compute

$$\operatorname{grad} V(r) = V'(r) \cdot \operatorname{grad} r = mf(r) \cdot \frac{\mathbf{r}}{r}.$$

Remark 1.11 In the Newtonian case, with $f(r) = \mu/r^2$, the usual normalisation convention is to take $r_0 = \infty$, i.e.

$$V(r) = m \int_{\infty}^{r} \frac{\mu}{\rho^2} \,\mathrm{d}\rho = -\frac{m\mu}{r}.$$

Notes and references

The path-lifting property for coverings alluded to in Remark 1.4 is not difficult to show, but it requires a careful argument; see (Jänich, 2005, Chapter 9) or (McCleary, 2006, Chapter 8).

Kepler's second law was originally proved by Newton (1687) in Book 1, Proposition 1 of his *Principia*. A very useful guide to Newton's masterpiece is (Chandrasekhar, 1995).

Exercises

1.1 Let $\mathbb{R} \ni t \mapsto \mathbf{r}(t) \in \mathbb{R}^3 \setminus \{\mathbf{0}\}$ be a solution of the central force problem (CFP). In this exercise we investigate the symmetries of such a solution.

(a) Show that $t \mapsto \mathbf{r}(t+b)$ for $b \in \mathbb{R}$ and $t \mapsto \mathbf{r}(-t)$ are likewise solutions of (CFP), i.e. we have invariance under time translation and time reversal.

(b) Suppose we are in the centrally symmetric situation, i.e. the function f in (CFP) depends only on r rather than $\mathbf{r}$. Show that in this case we have invariance under **isometries**, i.e. distance-preserving maps: if $A\colon \mathbb{R}^3 \to \mathbb{R}^3$ is an isometry with $A(\mathbf{0}) = \mathbf{0}$, so that A may be regarded as an element of the orthogonal group $\mathrm{O}(3)$, then $t \mapsto A\mathbf{r}(t)$ is likewise a solution.

1.2 Verify the Graßmann identity (1.1).

Hint: Observe that both sides of the equation are linear in $\mathbf{a}, \mathbf{b}$ and $\mathbf{c}$. It therefore suffices to check equality when $\mathbf{a}, \mathbf{b}, \mathbf{c}$ are chosen from the three standard basis vectors of $\mathbb{R}^3$.

1.3 Verify that the map p in Remark 1.4 is a covering map in the sense described there. As index set one can take $\Lambda = \mathbb{Z}$. How does one have to choose the neighbourhood U of a given point in $\mathbb{R}^2 \setminus \{\mathbf{0}\}$?

1.4 Let $I \subset \mathbb{R}$ be an interval and $\boldsymbol{\alpha}\colon I \to \mathbb{R}^2 \setminus \{\mathbf{0}\}$ a C^k-curve for some $k \in \mathbb{N}$. We write

$$\boldsymbol{\alpha}(t) = (\alpha_1(t), \alpha_2(t)) = r(t)\,(\cos\theta(t), \sin\theta(t))$$

with $r(t) > 0$. For every $t \in I$, the angle $\theta(t)$ is determined by $\boldsymbol{\alpha}(t)$ up to adding integer multiples of 2π. The aim of this exercise is to give an alternative proof of the observation in Remark 1.4, i.e. that θ can be chosen as a C^k-function.

First of all, we observe that r is determined by $r(t) = |\boldsymbol{\alpha}(t)|$ and hence is of class C^k, since $\boldsymbol{\alpha} \neq \mathbf{0}$. Thus, by passing to the curve $\boldsymbol{\alpha}/r$ we may assume without loss of generality that $r = 1$.

(a) (Uniqueness) Suppose the planar curve $\boldsymbol{\alpha} = (\alpha_1, \alpha_2)$ has been written as $\boldsymbol{\alpha} = (\cos\theta, \sin\theta)$ with a C^k-function θ. Let $t_0 \in I$ and $\theta_0 = \theta(t_0)$. Show that

$$\theta(t) = \theta_0 + \int_{t_0}^{t} (\alpha_1(s)\dot{\alpha}_2(s) - \dot{\alpha}_1(s)\alpha_2(s))\,\mathrm{d}s.$$

(b) (Existence) Let $\boldsymbol{\alpha} = (\alpha_1, \alpha_2)$ and $t_0 \in I$ be given. Choose θ_0 such that $\boldsymbol{\alpha}(t_0) = (\cos\theta_0, \sin\theta_0)$, and define θ via the equation in (a). Set $(\beta_1, \beta_2) := (\cos\theta, \sin\theta)$. Show that (α_1, α_2) and (β_1, β_2) are solutions of one and the same linear system of differential equations, with equal initial values at $t = t_0$. Conclude with the uniqueness statement in the Picard–Lindelöf theorem that $(\alpha_1, \alpha_2) = (\beta_1, \beta_2)$. Verify that the assumptions of that theorem are indeed satisfied.

1.5 In this exercise we want to give an alternative proof of the area formula for planar sets D of the form

$$D = \{s\alpha(t)\colon\ t \in [t_0, t_1],\ s \in [0, 1]\},$$

where $\alpha = r(\cos\theta, \sin\theta)\colon\ [t_0, t_1] \to \mathbb{R}^2 \setminus \{\mathbf{0}\}$ is a C^1-curve with $\dot\theta > 0$ on $[t_0, t_1]$ and $\theta(t_1) - \theta(t_0) < 2\pi$, cf. Proposition 1.5.

(a) Show that the exterior normal vector $\mathbf{n}(t)$ to D in the boundary point $\alpha(t) \in \partial D$ is given by

$$\mathbf{n}(t) = \frac{(\dot\alpha_2(t), -\dot\alpha_1(t))}{|\dot\alpha(t)|}.$$

(b) Apply the divergence theorem (a.k.a. Gauß's theorem)

$$\int_D \operatorname{div}\mathbf{v}\, \mathrm{d}x\, \mathrm{d}y = \int_{\partial D} \langle \mathbf{v}, \mathbf{n}\rangle\, \mathrm{d}s$$

to the vector field $\mathbf{v}(x, y) = (x, y)$ in order to derive the area formula from Proposition 1.5. What is the contribution of the line segments in ∂D to the boundary integral?

1.6 (a) Let $t \mapsto \mathbf{r}(t) \in \mathbb{R}^3 \setminus \{\mathbf{0}\}$ be a C^2-map. Set $\mathbf{v} := \dot{\mathbf{r}}$ and $\mathbf{c} := \mathbf{r} \times \dot{\mathbf{r}}$. Show that

$$v^2 = \dot r^2 + \frac{c^2}{r^2}.$$

(b) Now assume that $\mathbf{r}$ is a solution of the central force problem. So we may take $\mathbf{r}$ to be a planar curve and write $\mathbf{r} = r(\cos\theta, \sin\theta)$. Set $\mathbf{a} = \ddot{\mathbf{r}}$. Show that

$$a = \left|\frac{c^2}{r^3} - \ddot r\right|.$$

1.7 In this exercise we wish to derive a special case of Kepler's third law, see Theorem 3.7. Let $a \in \mathbb{R}^+$ and $\omega \in \mathbb{R}$. Show that

$$\mathbf{r}(t) := a(\cos\omega t, \sin\omega t)$$

is a solution of the central force problem $\ddot{\mathbf{r}} = -\mathbf{r}/r^3$ (corresponding to the Newtonian law of gravitation $F \propto r^{-2}$) if and only if $|\omega| = 1/a^{3/2}$. What is the relation between the minimal period p (i.e. the time for one full rotation) and the radius a?

1.8 Verify the formula $\operatorname{grad} r = \mathbf{r}/r$ used in Example 1.10, and interpret this formula geometrically.

2

Conic sections

Wenn sie erst den Kegel sieht, *muß sie glücklich sein.*

Thomas Bernhard, *Korrektur*

As we shall see in the next chapter, when we specialise in the central force problem to the Newtonian law of gravitational attraction, the solution curve will be an ellipse, a parabola, or a hyperbola. In the present chapter I give a bare bones introduction to the theory of these planar curves.

2.1 Ellipses

We take the following *gardener's construction* as the definition of an ellipse, see Figure 2.1, and then derive five other equivalent characterisations. The distance between two points $P, Q \in \mathbb{R}^2$ will be denoted by $|PQ|$.

Definition 2.1 Let F_1, F_2 be two points in $\mathbb{R}^2$ (possibly $F_1 = F_2$), and choose a real number $a > \frac{1}{2}|F_1F_2|$. The **ellipse** with **foci** F_1, F_2 and **semi-major axis** a is the set

$$\mathcal{E} := \{P \in \mathbb{R}^2 \colon\ |PF_1| + |PF_2| = 2a\}.$$

If you wish to lay out an elliptic flower bed in your garden, proceed as follows. Drive pegs into the ground at F_1 and F_2, take a piece of string of length greater than $|F_1F_2|$, tie one end each to the pegs, and draw an ellipse by moving a marker P around the two pegs while keeping the string stretched tight with the marker.

The midpoint Z of the two foci is called the **centre** of the ellipse. If $F_1 = F_2 = Z$, then $\mathcal{E}$ is a circle of radius a about Z. If $F_1 \neq F_2$, let P_1 be the point on

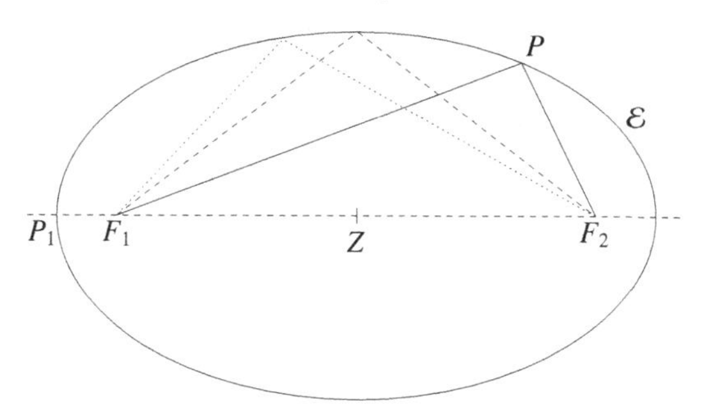

Figure 2.1 The gardener's construction of an ellipse.

$\mathcal{E}$ lying on the line through the foci on the side of F_1, as shown in Figure 2.1. Set $d_i = |F_iP_1|$, $i = 1, 2$. Then

$$|ZP_1| = \frac{d_2 - d_1}{2} + d_1 = \frac{d_1 + d_2}{2} = a.$$

The following proposition gives a description of ellipses in terms of a vector equation, so we use vector notation for points in $\mathbb{R}^2$.

Proposition 2.2 *The ellipse with foci* $\mathbf{A}, \mathbf{0}$ *(the origin of* $\mathbb{R}^2$*) and semi-major axis* a *is given by the equation*

$$\boxed{r + \langle \mathbf{e}, \mathbf{r} \rangle = a(1 - e^2),} \tag{2.1}$$

where $\mathbf{e} := -\mathbf{A}/2a$.

Observe that we must have $2a > A$, hence $0 \le e < 1$. This number e is called the **eccentricity** of the ellipse. The vector $\mathbf{e}$ will be referred to as the **eccentricity vector**.

Proof of Proposition 2.2 For the given foci, Definition 2.1 translates into

$$r + |\mathbf{A} - \mathbf{r}| = 2a.$$

Because of

$$|\mathbf{A} - \mathbf{r}| \ge r - A > r - 2a,$$

the equation $|\mathbf{A} - \mathbf{r}| = r - 2a$ cannot be satisfied, so the above equation is equivalent to the one obtained by squaring:

$$|\mathbf{A} - \mathbf{r}|^2 = (2a - r)^2.$$

We compute further:

$$\begin{aligned} A^2 + r^2 - 2\langle \mathbf{A}, \mathbf{r} \rangle &= 4a^2 + r^2 - 4ar \\ r - \langle \mathbf{A}/2a, \mathbf{r} \rangle &= a - A^2/4a \\ r + \langle \mathbf{e}, \mathbf{r} \rangle &= a(1 - e^2). \end{aligned}$$

□

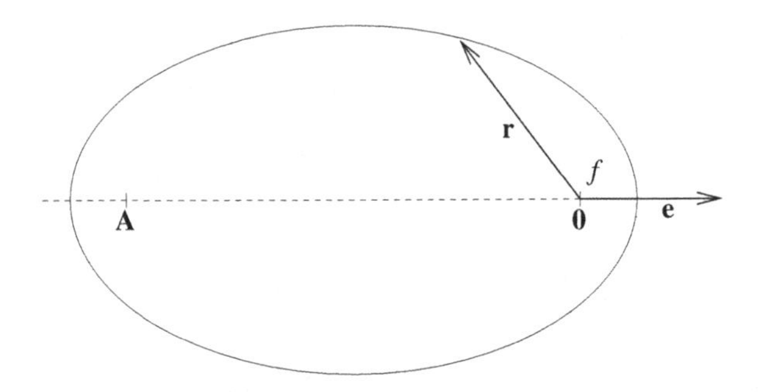

Figure 2.2 Polar coordinates relative to $\mathbf{e}$.

We now rewrite this equation in polar coordinates. Let $f = f(\mathbf{r})$ be the angle between $\mathbf{e}$ and $\mathbf{r}$, see Figure 2.2.[1] We then immediately obtain the following corollary from Proposition 2.2.

Corollary 2.3 *In polar coordinates* (r, f) *relative to the axis defined by the eccentricity vector* $\mathbf{e}$, *the equation*

$$\boxed{r = \frac{a(1 - e^2)}{1 + e\cos f}}$$

describes the ellipse with one focus at the origin and semi-major axis a. □

From this description in terms of polar coordinates we can easily read off a further geometric interpretation of the eccentricity e. For f equal to an even or odd integer multiple of π, the radius r attains its extremal values

$$r_{\min} = \frac{a(1 - e^2)}{1 + e} = a(1 - e) \quad \text{and} \quad r_{\max} = \frac{a(1 - e^2)}{1 - e} = a(1 + e),$$

respectively. It follows that the distance from the centre Z of the ellipse $\mathcal{E}$ to one of the foci equals ea, see Figure 2.3. In particular, we see that $\mathcal{E}$ is determined up to isometry by a and e.

Write b for the distance between Z and one of the two points on $\mathcal{E}$ symmetric

[1] The slightly unusual notation f for an angle is due to historical conventions. No confusion should arise with the use of f to describe a general central force field in Chapter 1.

with respect to the two foci; this distance is called the **semi-minor axis**. From the right-angled triangle in Figure 2.3 we read off

$$b = a\sqrt{1 - e^2}.$$

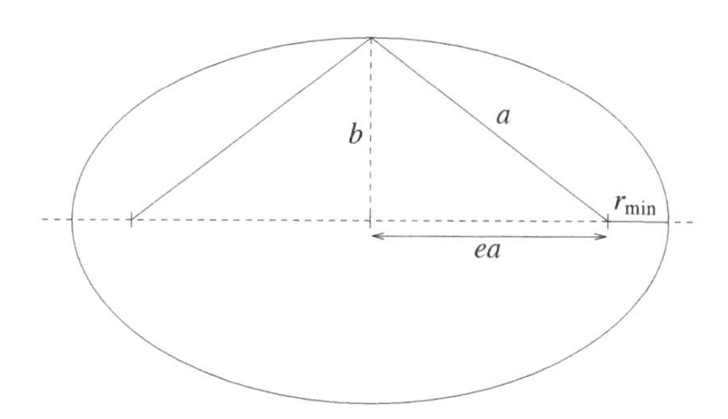

Figure 2.3 Eccentricity and the semi-axes.

For our purposes, the descriptions of an ellipse by a vector equation as in Proposition 2.2 or by an equation in polar coordinates as in Corollary 2.3 are the most relevant. The following alternative descriptions are included for completeness.

Proposition 2.4 *Let ℓ be a line in $\mathbb{R}^2$ and $F \in \mathbb{R}^2$ a point not on ℓ. Choose a real number $0 < e < 1$. Then the set of points $P \in \mathbb{R}^2$ satisfying*

$$|PF| = e\,\mathrm{dist}(P, \ell) \tag{2.2}$$

is the ellipse with one focus at F, eccentricity vector $\mathbf{e}$ equal to the vector orthogonal to ℓ of length e, pointing from F towards ℓ, and semi-major axis

$$a = \frac{e\,\mathrm{dist}(F, \ell)}{1 - e^2}.$$

The line ℓ is called the **directrix** *of this ellipse, see Figure 2.4.*

Proof We may assume that F lies at the origin. Then $|PF| = r$. Condition (2.2) implies that P and F have to lie on the same side of ℓ. Hence

$$\begin{aligned} \mathrm{dist}(P, \ell) &= \mathrm{dist}(F, \ell) - \langle \mathbf{e}/e, \mathbf{r} \rangle \\ &= \frac{1}{e}(a(1 - e^2) - \langle \mathbf{e}, \mathbf{r} \rangle). \end{aligned}$$

So (2.2) translates directly into (2.1).

Conversely, let $\mathcal{E}$ be the ellipse given by (2.1). Define ℓ to be the line orthogonal to $\mathbf{e}$ at distance $a(1 - e^2)/e$ from $F = \mathbf{0}$ in the direction of $\mathbf{e}$. Observe

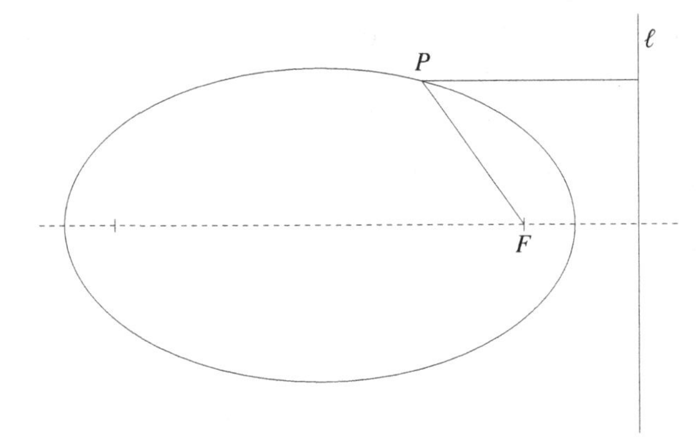

Figure 2.4 The directrix.

from (2.1) that the point P_0 on $\mathcal{E}$ farthest from F in the direction of $\mathbf{e}$ (i.e. with $\langle \mathbf{e}, \mathbf{r} \rangle$ maximal) is actually the point closest to F (i.e. with r minimal), see also Figures 2.2 and 2.3. Hence

$$|FP_0| = r_{\min} = a(1-e) < a(1-e)\frac{1+e}{e} = \frac{a(1-e^2)}{e} = \operatorname{dist}(F, \ell).$$

This means that all points on $\mathcal{E}$ lie on the same side of ℓ as F. Then the same computation as above shows that (2.1) translates into (2.2). □

Proposition 2.5 *Let $\mathcal{E}$ be an ellipse with semi-major axis a and eccentricity e. Up to an isometry, we may assume that $\mathcal{E}$ is centred at the origin, and the foci are*

$$F_1 = (-ea, 0) \text{ and } F_2 = (ea, 0).$$

Then

$$\mathcal{E} = \left\{ (x, y) \in \mathbb{R}^2 : \frac{x^2}{a^2} + \frac{y^2}{b^2} = 1 \right\},$$

where $b = a\sqrt{1-e^2}$.

Proof For $e = 0$ the statement is clear. For $0 < e < 1$, the preceding proposition says that $\mathcal{E}$ can equivalently be described by (2.2), where $F = F_2 = (ea, 0)$ and ℓ is the vertical line given by

$$x = ea + \frac{a(1-e^2)}{e} = \frac{a}{e}.$$

Written out in cartesian coordinates, equation (2.2) then becomes

$$(x - ea)^2 + y^2 = e^2\left(x - \frac{a}{e}\right)^2.$$

A straightforward computation transforms this into the quadratic equation in the proposition. □

Finally, there is a characterisation of ellipses that justifies calling them a *conic section* (or *conic*, for short), see Exercise 2.1.

2.2 Hyperbolas

As the definition of hyperbolas we take the following analogue of the gardener's construction, see Figure 2.5.

Definition 2.6 Let F_1, F_2 be two points in $\mathbb{R}^2$, and let $a \in \mathbb{R}^+$ be a real number with $2a < |F_1F_2|$. The **hyperbola** with **foci** F_1, F_2 and **real semi-axis** a is the set

$$\mathcal{H} := \{P \in \mathbb{R}^2 \colon\ |PF_2| - |PF_1| = \pm 2a\}.$$

Take a rod of length L and a piece of string of length l with $L - l = 2a$. Regard F_2 as a pivot about which the rod turns. Tie one end of the string to the free end Q of the rod, and the other end to F_1. Pull the string tight along the rod, so that it lies straight along the rod from Q to some point P, and then forms a straight line from P to F_1. We then have

$$|PF_2| - |PF_1| = |QF_2| - |QP| - |PF_1| = L - l = 2a.$$

In this way we obtain one of the two **branches** $\mathcal{H}^\pm$ of the hyperbola.[2]

Here is the analogue of Proposition 2.2.

Proposition 2.7 *The hyperbola with foci* $\mathbf{0}, \mathbf{A}$ *and real semi-axis* a *is given by the equation*

$$\boxed{r \pm \langle \mathbf{e}, \mathbf{r}\rangle = \pm a(e^2 - 1),} \tag{2.3}$$

where $\mathbf{e} := \mathbf{A}/2a$.

As before, we speak of the **eccentricity vector e** and the **eccentricity** e. Notice that for hyperbolas we have $e > 1$.

[2] More precisely, we obtain those points P on the hyperbola that, in addition, satisfy $|PF_1| \leq l$ or, equivalently, $|PF_2| \leq L$.

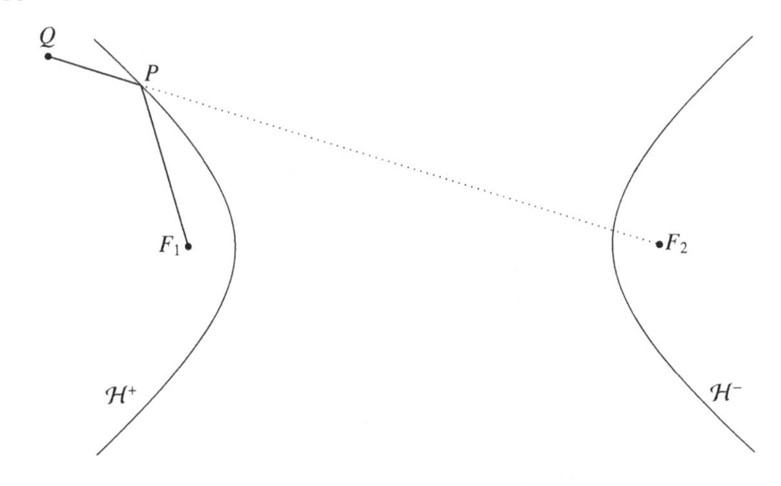

Figure 2.5 The gardener's construction of a hyperbola.

Proof of Proposition 2.7 Definition 2.6 translates into

$$|\mathbf{A} - \mathbf{r}| - r = \pm 2a.$$

We consider the case of a plus sign on the right-hand side. Because of $r + 2a > 0$, the equation is then equivalent to

$$|\mathbf{A} - \mathbf{r}|^2 = (r + 2a)^2.$$

We compute further:

$$\begin{aligned} A^2 + r^2 - 2\langle \mathbf{A}, \mathbf{r} \rangle &= r^2 + 4a^2 + 4ar \\ r + \langle \mathbf{A}/2a, \mathbf{r} \rangle &= A^2/4a - a \\ r + \langle \mathbf{e}, \mathbf{r} \rangle &= a(e^2 - 1). \end{aligned}$$

The case of a minus sign on the right-hand side is analogous, see Exercise 2.3. □

Definition 2.8 The part of the hyperbola described by (2.3) with plus signs, i.e. those points closer to **0** than to **A**, will be called the **principal branch** of the hyperbola with respect to the focus **0**; see Figure 2.6.

For attractive Newtonian forces with centre at **0**, only this principal branch will be relevant. The secondary branch would describe orbits in a repelling force field centred at **0**.

The following analogue of Corollary 2.3 is immediately apparent.

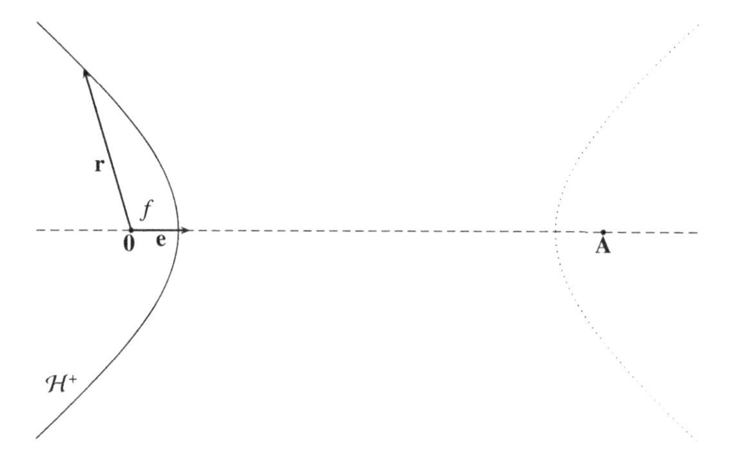

Figure 2.6 The principal branch of a hyperbola and polar coordinates.

Corollary 2.9 *In polar coordinates (r, f) relative to the axis defined by the eccentricity vector* **e**, *the equation for the principal branch relative to the focus* **0** *and real semi-axis a is*

$$\boxed{r = \frac{a(e^2 - 1)}{1 + e\cos f}.}$$

Here the angle f is allowed to vary in

$$(-\arccos(-1/e), \arccos(-1/e)) \subset (-\pi, \pi),$$

see Figure 2.7. □

From the description in terms of polar coordinates we see that for points on the principal branch, the distance r reaches its minimal value

$$r_{\min} = \frac{a(e^2 - 1)}{1 + e} = a(e - 1)$$

for $f = 0$.

Here is the description of the principal branch of a hyperbola in terms of a directrix, analogous to Proposition 2.4. The proof is left as an exercise.

Proposition 2.10 *Let ℓ be a line in $\mathbb{R}^2$ and $F \in \mathbb{R}^2$ a point not on ℓ. Choose a real number $e > 1$. Then the set of points satisfying*

$$|PF| = e\,\mathrm{dist}(P, \ell) \tag{2.4}$$

is the hyperbola with one focus at F, eccentricity vector **e** *equal to the vector*

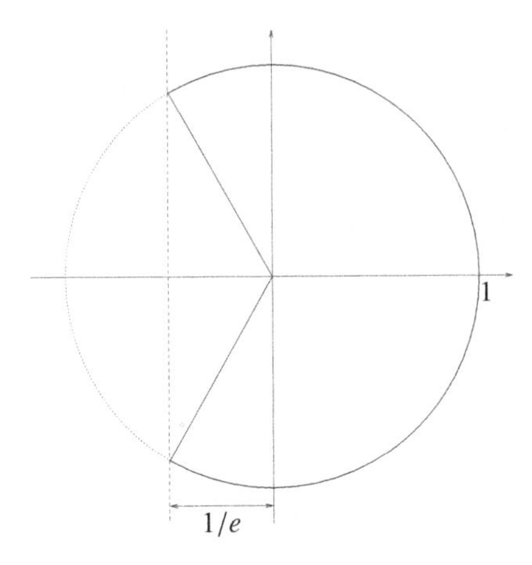

Figure 2.7 The range of the angle f.

orthogonal to ℓ of length e, pointing from F towards ℓ, and real semi-axis

$$a = \frac{e \operatorname{dist}(F, \ell)}{e^2 - 1}.$$

The principal branch relative to F is the set of those points P that lie on the same side of ℓ as F. We call ℓ the **directrix** *of the principal branch.* □

The analogue of Proposition 2.5 is given in Exercise 2.5.

2.3 Parabolas

The description of ellipses and hyperbolas in terms of a directrix obviously leaves open the case with eccentricity $e = 1$. In this case, that description can also be seen as a gardener's construction, see Figure 2.8.

Definition 2.11 Let ℓ be a line in $\mathbb{R}^2$ and $F \in \mathbb{R}^2$ a point not on ℓ. The **parabola** with **focus** F and **directrix** ℓ is the set

$$\mathcal{P} := \{P \in \mathbb{R}^2 \colon\ |PF| = \operatorname{dist}(P, \ell)\}.$$

The following analogue of Propositions 2.2 and 2.7 is proved in Figure 2.9.

Proposition 2.12 *The parabola with focus* $\mathbf{0}$ *and directrix ℓ at distance d from* $\mathbf{0}$ *is given by the equation*

$$\boxed{r + \langle \mathbf{e}, \mathbf{r} \rangle = d,}$$

where $\mathbf{e}$ *is the unit vector orthogonal to ℓ, directed from* $\mathbf{0}$ *to ℓ.* □

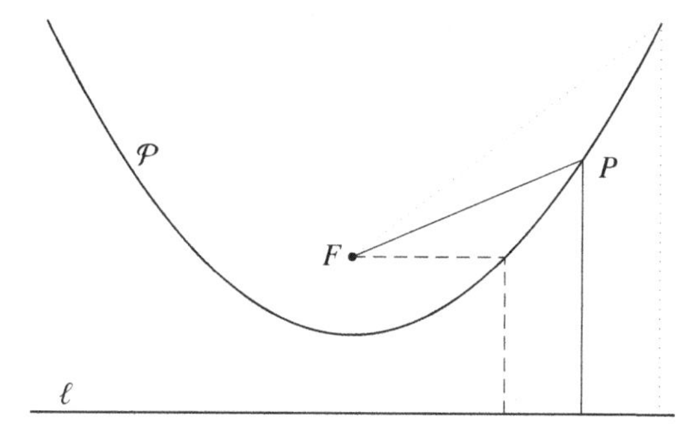

Figure 2.8 The gardener's construction of a parabola.

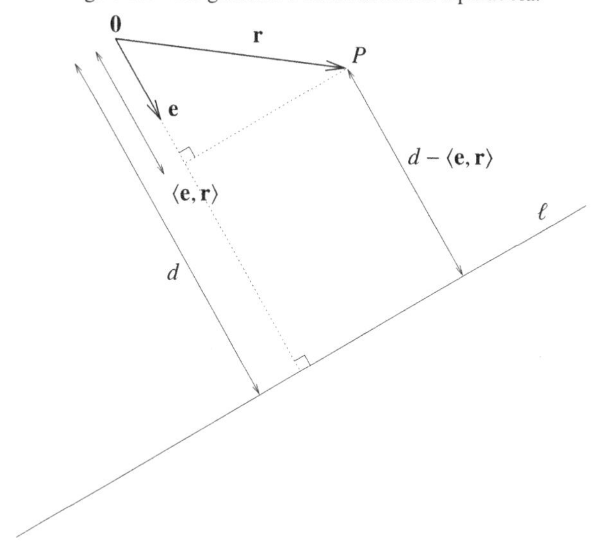

Figure 2.9 Proof of Proposition 2.12.

Again we can easily reformulate this equation for a parabola in terms of polar coordinates.

Corollary 2.13 *In polar coordinates (r, f) relative to the axis defined by the eccentricity vector* $\mathbf{e}$*, the equation for the parabola with focus at the origin and directrix at distance d from* $\mathbf{0}$ *in the direction of* $\mathbf{e}$ *is*

$$\boxed{r = \frac{d}{1 + \cos f},}$$

where f is allowed to vary in the open interval $(-\pi, \pi)$. □

The next proposition is the analogue of Proposition 2.5 and Exercise 2.5.

Proposition 2.14 *The parabola $\mathcal{P}$ with focus $F = (0, p)$ and directrix $\ell = \{y = -p\}$ is given by*

$$\mathcal{P} = \{(x, y) \in \mathbb{R}^2 : x^2 = 4py\}.$$

Proof With $|PF|^2 = x^2 + (y - p)^2$ and $\text{dist}^2(P, \ell) = (y + p)^2$ for $P = (x, y)$, Definition 2.11 translates into the equation given in the proposition. □

Here, finally, is the description of parabolas as conic sections.

Proposition 2.15 *The curve $\mathcal{P}$ given by intersecting the cone*

$$\{(x, y, z) \in \mathbb{R}^3 : x^2 + y^2 - z^2 = 0\}$$

with the affine plane

$$\{(x, y, z) \in \mathbb{R}^3 : z = y + c\},$$

where $c > 0$, is a parabola.

Proof We substitute $z = y + c$ into the equation for the cone to obtain

$$\mathcal{P} = \left\{\left(x, \frac{x^2}{2c} - \frac{c}{2}, \frac{x^2}{2c} + \frac{c}{2}\right) : x \in \mathbb{R}\right\}.$$

By a translation $(x, y, z) \mapsto (x, y + c/2, z - c/2)$, followed by a rotation about the x-axis, we can map $\mathcal{P}$ isometrically to the parabola

$$\{(x, x^2/\sqrt{2}c, 0) : x \in \mathbb{R}\}.$$

□

2.4 Summary

Here is a unified description of the three types of conic sections.

Corollary 2.16 *Let ℓ be a line in $\mathbb{R}^2$ and $F \in \mathbb{R}^2$ a point not on ℓ. Let e be a positive real number and* **e** *the vector of length e pointing orthogonally from F to ℓ. Then the set of points $P \in \mathbb{R}^2$ on the same side of ℓ as F, satisfying*

$$|PF| = e\,\text{dist}(P, \ell),$$

is an ellipse for $0 < e < 1$, a parabola for $e = 1$, and the principal branch of a hyperbola for $e > 1$, each with one focus at F. The equation

$$r = \frac{e\,\text{dist}(F, \ell)}{1 + e \cos f}$$

gives a description of this conic in polar coordinates (r, f) relative to $F = \mathbf{0}$ and **e**. □

Notes and references

The first systematic account of conic sections was given by Apollonius of Perga (third century BC). For a concise modern introduction to conic sections see Chapter 4 of (Knörrer, 1996).

Exercises

2.1 Consider the curve $\mathcal{E}$ given by intersecting the cone

$$\{(x, y, z) \in \mathbb{R}^3 : \ x^2 + y^2 - z^2 = 0\}$$

with the affine plane

$$E := \{(x, y, z) \in \mathbb{R}^3 : \ z = mx + c\},$$

where c, m are real numbers with $c > 0$ and $0 \leq m < 1$. Show that there is an isometry $E \to \mathbb{R}^2 \times \{0\} \subset \mathbb{R}^3$ that sends $\mathcal{E}$ to an ellipse in standard form $\{x^2/a^2 + y^2/b^2 = 1\}$. Determine the semi-axes a and b explicitly.

2.2 Verify directly that the characterisations of ellipses via the gardener's construction and in terms of a quadratic equation as in Proposition 2.5 are equivalent.

2.3 Verify the second case in Proposition 2.7: the equation

$$|\mathbf{A} - \mathbf{r}| - r = -2a$$

is equivalent to

$$r - \langle \mathbf{e}, \mathbf{r} \rangle = -a(e^2 - 1).$$

2.4 Prove Proposition 2.10.

2.5 (a) Let $a, b \in \mathbb{R}^+$ be given. Define $e \in \mathbb{R}^+$, $e > 1$, by the equation

$$b^2 = a^2(e^2 - 1).$$

Sketch the set

$$\mathcal{H} := \left\{(x, y) \in \mathbb{R}^2 : \ \frac{x^2}{a^2} - \frac{y^2}{b^2} = 1\right\}.$$

Let F_1 be the point $(-ae, 0) \in \mathbb{R}^2$ and F_2 the point $(ae, 0)$. Show that

$$\mathcal{H} = \{P \in \mathbb{R}^2 : \ |F_1P| - |F_2P| = \pm 2a\},$$

i.e. $\mathcal{H}$ is a hyperbola.

(b) Compute the distance of a point $(x, y) \in \mathcal{H} \cap (\mathbb{R}^+ \times \mathbb{R}^+)$ from the line $\{y = bx/a\}$. Show that this distance goes to zero for $x \to \infty$, and that $\mathcal{H} \cap (\mathbb{R}^+ \times \mathbb{R}^+)$ lies on one side of that line. The line is called an **asymptote** of the hyperbola.

2.6 What is the hyperbolic analogue of the statements in Exercise 2.1 and Proposition 2.15?

2.7 We consider the principal branch $\mathcal{H}^+$ of a hyperbola in the xy-plane as in Figure 2.6, where the cartesian coordinates are chosen in such a way that the corresponding focus lies at the origin and the eccentricity vector $\mathbf{e}$ points in the positive x-direction. Let a be the real semi-axis of the hyperbola. Show the following.

(a) $\mathcal{H}^+$ is described by the equation

$$\sqrt{x^2 + y^2} + ex = a(e^2 - 1).$$

(b) This equation is equivalent to

$$(x - ea)^2 - \frac{y^2}{e^2 - 1} = a^2, \quad x - ea \leq -a.$$

(c) A parametrisation of $\mathcal{H}^+$ is given by

$$u \longmapsto a(e - \cosh u, \sqrt{e^2 - 1} \sinh u), \quad u \in \mathbb{R}.$$

3

The Kepler problem

> Mr. Hooke sent, in his next letter [to Isaac Newton], the whole of his Hypothesis, *scil.* that the gravitation was reciprocall to the square of the distance: which is the whole coelastiall theory, concerning which Mr. Newton haz made a demonstration, not at all owning he receiv'd the first Intimation of it from Mr. Hooke. [...] This is the greatest Discovery in Nature that ever was since the World's Creation. It never was so much as hinted by any man before. I wish he had writt plainer, and afforded a little more paper.
>
> John Aubrey, *Brief Lives*

We now specialise to what is called the Newtonian law of attraction in the central force problem, notwithstanding the anticipation of the inverse square law by other authors (see the notes and references section).

The Kepler problem Find integral curves of the differential equation

$$\ddot{\mathbf{r}} = -\frac{\mu}{r^2} \cdot \frac{\mathbf{r}}{r}, \tag{K}$$

where $\mu > 0$ is a given real constant.

In Section 3.1 we prove Kepler's first law, which states that solutions of (K) with non-vanishing angular momentum trace out conic sections with one focus at the origin. In Section 3.2 we relate the eccentricity of that conic section to the energy of the solution. Section 3.3 contains a proof of Kepler's third law about the relation between the minimal period of an elliptic solution and the semi-major axis of the ellipse.

3.1 Kepler's first law

Theorem 3.1 (Kepler's first law) *Let* $\mathbf{r}\colon I \to \mathbb{R}^3 \setminus \{\mathbf{0}\}$ *be a solution of the Kepler problem* (K) *with non-vanishing angular momentum* $\mathbf{c}$*. Then* $\mathbf{r}$ *lies on a conic section (ellipse, parabola or hyperbola) with one focus at the origin.*

Proof Recall equation (1.2):

$$\frac{\mathrm{d}}{\mathrm{d}t}\left(\frac{\mathbf{r}}{r}\right) = \frac{\mathbf{c}\times\mathbf{r}}{r^3}.$$

Hence, for a solution of (K) we have

$$\frac{\mathrm{d}}{\mathrm{d}t}\left(\mu\frac{\mathbf{r}}{r}\right) = -\mathbf{c}\times\ddot{\mathbf{r}} = -\frac{\mathrm{d}}{\mathrm{d}t}(\mathbf{c}\times\dot{\mathbf{r}}),$$

since $\mathbf{c}$ is constant by Proposition 1.3. Integrating this equation, we get

$$\mu\left(\frac{\mathbf{r}}{r}+\mathbf{e}\right) = -\mathbf{c}\times\dot{\mathbf{r}}, \tag{3.1}$$

where we have written the constant of integration as $\mu\mathbf{e}$.

- For $\mathbf{c} = \mathbf{0}$ we have $\mathbf{e} = -\mathbf{r}/r$, i.e. the vector $\mathbf{e}$ is a unit vector along the line through $\mathbf{0}$ on which the motion takes place.[1]
- For $\mathbf{c} \neq \mathbf{0}$, taking the inner product of (3.1) with $\mathbf{c}$ (and observing that $\langle\mathbf{c},\mathbf{r}\rangle = 0$) yields $\langle\mathbf{c},\mathbf{e}\rangle = 0$, i.e. the vector $\mathbf{e}$ lies in the plane of motion.

Now, taking the inner product of (3.1) with $\mathbf{r}$, we find

$$\begin{aligned}\mu(r+\langle\mathbf{e},\mathbf{r}\rangle) &= -\langle\mathbf{c}\times\dot{\mathbf{r}},\mathbf{r}\rangle\\ &= -\langle\mathbf{c},\dot{\mathbf{r}}\times\mathbf{r}\rangle\\ &= c^2,\end{aligned}$$

where we have used the identity

$$\langle\mathbf{a}\times\mathbf{b},\mathbf{c}\rangle = \langle\mathbf{a},\mathbf{b}\times\mathbf{c}\rangle \tag{3.2}$$

for arbitrary vectors $\mathbf{a},\mathbf{b},\mathbf{c}\in\mathbb{R}^3$ (see Exercise 3.1). Hence,

$$r+\langle\mathbf{e},\mathbf{r}\rangle = c^2/\mu, \tag{3.3}$$

which, for $c \neq 0$, is the equation of a conic section by Propositions 2.2, 2.7 and 2.12. □

Remarks 3.2 (1) In the case of a hyperbola, the sign in equation (3.3) corresponds to the principal branch of the hyperbola relative to the focus $\mathbf{0}$.

[1] For more on this case see Exercise 4.7.

(2) By comparing (3.3) with (2.1) and (2.3), we find

$$\boxed{\frac{c^2}{\mu} = a|1 - e^2| \text{ for } c \neq 0, e \neq 1.} \tag{3.4}$$

(3) The computation leading to (3.1) shows that for any solution $\mathbf{r}$ of (K), the vector $\dot{\mathbf{r}} \times \mathbf{c}/\mu - \mathbf{r}/r$ is a constant of motion, and by (3.3) it equals the eccentricity vector. In the physics literature this constant of motion is usually called the Laplace–Runge–Lenz vector.

Corollary 3.3 *The solutions of* (K) *with $c \neq 0$ are defined for all $t \in \mathbb{R}$, and the whole conic section is traversed in one direction – infinitely often in the elliptic case – as t goes from $-\infty$ to $+\infty$.*

Proof We regard (K) as a system of first order:

$$\left.\begin{aligned} \dot{\mathbf{r}} &= \mathbf{v} \\ \dot{\mathbf{v}} &= -\frac{\mu}{r^2} \cdot \frac{\mathbf{r}}{r}. \end{aligned}\right\} \tag{K$'$}$$

By the Picard–Lindelöf theorem, the initial value problem

$$\left.\begin{aligned} \mathbf{r}(0) &= \mathbf{r}_0 \\ \mathbf{v}(0) &= \mathbf{v}_0 \end{aligned}\right\}$$

has a unique solution, which is defined on a maximal interval (α, ω) containing 0. We want to show that $\omega = \infty$; the proof that $\alpha = -\infty$ is analogous.

There are two reasons why we might have $\omega < \infty$.

(i) There is a sequence $t_\nu \nearrow \omega$ with $\mathbf{r}(t_\nu) \to \mathbf{0}$, i.e. the solution approaches the boundary of the domain of definition $(\mathbb{R}^3 \setminus \{\mathbf{0}\}) \times \mathbb{R}^3$ of the right-hand side of (K$'$).

(ii) There is a sequence $t_\nu \nearrow \omega$ with $|\mathbf{r}(t_\nu)| + |\mathbf{v}(t_\nu)| \to \infty$, i.e. the solution 'blows up' in finite time.

That these are indeed the only potential reasons for $\omega < \infty$ is a standard fact in the theory of differential equations, but for the sake of completeness I include the argument. Arguing by contradiction, we suppose that $\omega < \infty$, but neither (i) nor (ii) occurs. This implies that for $t \to \omega$ the pair $(\mathbf{r}(t), \mathbf{v}(t))$ stays inside a compact region where the right-hand side of (K$'$) satisfies a global Lipschitz condition in $(\mathbf{r}, \mathbf{v})$. Moreover, there is a sequence $t_\nu \nearrow \omega$ such that $(\mathbf{r}(t_\nu), \mathbf{v}(t_\nu))$ converges to a limit point $(\mathbf{r}_\omega, \mathbf{v}_\omega) \in (\mathbb{R}^3 \setminus \{\mathbf{0}\}) \times \mathbb{R}^3$. Again by Picard–Lindelöf, there is a neighbourhood U of $(\mathbf{r}_\omega, \mathbf{v}_\omega)$ and a constant $\delta > 0$ such that solutions of (K$'$) starting in U are defined on the interval $(-\delta, \delta)$. Taking as our initial value $(\mathbf{r}(t_\nu), \mathbf{v}(t_\nu))$ with ν sufficiently large, so that the initial value lies in U

and $\omega - t_v < \delta/2$, say, we can continue the solution beyond $t = \omega$, which is a contradiction.

Thus, it remains to show that neither (i) nor (ii) can occur. By Theorem 3.1 and the results in Chapter 2 we know that the polar coordinates $(r(t), f(t))$ of $\mathbf{r}(t)$ with respect to the centre $\mathbf{0}$ and axis $\mathbf{e}$ are related by

$$r = \frac{c^2/\mu}{1 + e\cos f}.$$

This implies

$$r \geq \frac{c^2/\mu}{1 + e},$$

which excludes possibility (i). With (K) we infer

$$|\ddot{\mathbf{r}}| \leq \frac{\mu^3(1+e)^2}{c^4}.$$

Then, from

$$\dot{\mathbf{r}}(t) = \dot{\mathbf{r}}(0) + \int_0^t \ddot{\mathbf{r}}(s)\,\mathrm{d}s$$

we deduce, for $t \geq 0$,

$$\begin{aligned} |\dot{\mathbf{r}}(t)| &\leq |\dot{\mathbf{r}}(0)| + \int_0^t |\ddot{\mathbf{r}}(s)|\,\mathrm{d}s \\ &\leq |\dot{\mathbf{r}}(0)| + \frac{\mu^3(1+e)^2}{c^4}\,t. \end{aligned}$$

If $\omega < \infty$, then $|\dot{\mathbf{r}}(t)|$ stays bounded in forward time. By repeating this integration argument, we find that $|\mathbf{r}(t)|$ likewise has an upper bound for $0 \leq t < \omega < \infty$. This precludes possibility (ii). So the solution of (K) is defined for all $t \in \mathbb{R}$.

In the proof of Kepler's second law (Theorem 1.6) we saw that $c = r^2\dot{f}$. In the elliptic case ($e < 1$) we have $r \leq r_{\max} = a(1+e)$, see page 12, hence

$$\dot{f} = \frac{c}{r^2} \geq \frac{c}{a^2(1+e)^2}. \tag{3.5}$$

This shows that the ellipse is traversed infinitely often with angular velocity $\dot{f}$ bounded from below.

For $e \geq 1$ we have

$$\dot{f} = \frac{c}{r^2} = \frac{\mu^2}{c^3}(1 + e\cos f)^2 > 0,$$

so f is monotone in t and must converge to a limit for $t \to \infty$. If this limit were smaller than $\arccos(-1/e)$, then $\dot{f}$ would be bounded from below by a positive

Table 3.1 *Relation between eccentricity e and energy h for $c \neq 0$.*

e	h	trajectory
< 1	< 0	ellipse
$= 1$	$= 0$	parabola
> 1	> 0	hyperbola

constant, an obvious contradiction. So f must indeed approach the limit angle for $t \to \infty$, and similarly for $t \to -\infty$. This means that the parabolic or hyperbolic curve is traversed completely. □

3.2 Eccentricity and energy

Let $\mathbf{r}\colon \mathbb{R} \to \mathbb{R}^3 \setminus \{\mathbf{0}\}$ be a solution of the Kepler problem. Example 1.10 and Remark 1.11 show that the Kepler problem is conservative with potential $V(r) = -m\mu/r$. Hence, by Proposition 1.8, the *energy per unit mass*

$$\boxed{h := \frac{1}{2}v^2(t) - \frac{\mu}{r(t)}} \tag{3.6}$$

is constant.

Proposition 3.4 *For solutions of the Kepler problem* (K) *we have*

$$\boxed{\mu^2(e^2 - 1) = 2hc^2.} \tag{3.7}$$

In particular, for $c \neq 0$ we have the relation between eccentricity and energy as listed in Table 3.1.

Proof We start from equation (3.1):

$$\mu\left(\frac{\mathbf{r}}{r} + \mathbf{e}\right) = -\mathbf{c} \times \dot{\mathbf{r}}.$$

Since $\mathbf{c}$ is orthogonal to $\dot{\mathbf{r}}$, we have $|\mathbf{c} \times \dot{\mathbf{r}}| = c|\dot{\mathbf{r}}| = cv$. Thus, taking the norm square on both sides of (3.1), we get

$$\mu^2(1 + e^2 + 2\langle \mathbf{r}/r, \mathbf{e}\rangle) = c^2v^2.$$

With equation (3.3) and the definition of h we can write this as

$$\mu^2\left(1 + e^2 + 2\frac{c^2}{\mu r} - 2\right) = c^2\left(2h + 2\frac{\mu}{r}\right),$$

from which equation (3.7) follows by collecting terms. □

Remark 3.5 By comparing equations (3.4) and (3.7), we see that

$$\boxed{a = \frac{\mu}{2|h|} \quad \text{for } c \neq 0, h \neq 0.} \tag{3.8}$$

This identity remains valid for $h < 0$ and $c = 0$ if $2a$ is interpreted as the maximal distance of the body from the origin, where $v = 0$ in (3.6). See also Exercise 4.7.

3.3 Kepler's third law

Let $\mathbf{r}$ be a solution of the Kepler problem with $c \neq 0$ and $h < 0$, i.e. an elliptic solution. Choose cartesian coordinates in $\mathbb{R}^3$ such that the angular momentum vector $\mathbf{c}$ points in the positive z-direction. Write ω for the angle from the positive x-axis to the eccentricity vector $\mathbf{e}$, and $f(t)$ for the angle from $\mathbf{e}$ to the position vector $\mathbf{r}(t)$, see Figure 3.1.[2] This angle f is traditionally called the **true anomaly**.

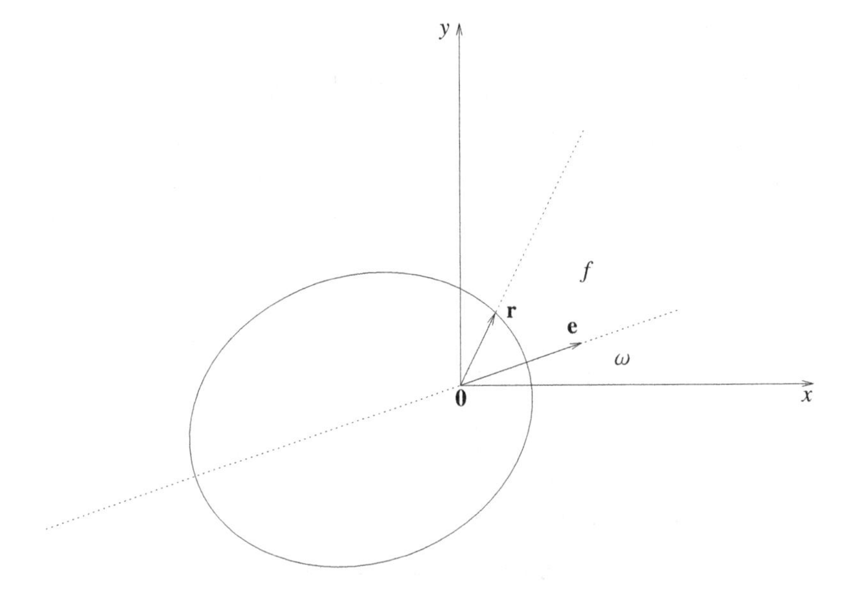

Figure 3.1 The angles ω and f.

Then the angle $\theta(t)$ from the positive x-axis to $\mathbf{r}(t)$ is given by $\theta(t) = \omega + f(t)$,

[2] In the circular case ($e = 0$) we choose an arbitrary reference direction to define f.

so that

$$\mathbf{r}(t) = r(t)\,(\cos\theta(t), \sin\theta(t), 0)$$

with

$$r(t) = \frac{a(1-e^2)}{1 + e\cos(\theta(t) - \omega)}.$$

By (3.5), the angular velocity $\dot{\theta}$ is bounded from below by a positive constant.

Lemma 3.6 *Define $p \in \mathbb{R}^+$ by $\theta(p) = \theta(0) + 2\pi$. Then p is a* **period** *of* $\mathbf{r}$, *i.e.*

$$\mathbf{r}(t+p) = \mathbf{r}(t) \;\; \textit{for all} \;\; t \in \mathbb{R},$$

and there is no smaller $p \in \mathbb{R}^+$ with this property.

Proof From the definition of p and the equations above for r and $\mathbf{r}$ we have $\mathbf{r}(p) = \mathbf{r}(0)$. Then, from

$$r^2(p)\dot{\theta}(p) = c = r^2(0)\dot{\theta}(0)$$

we deduce $\dot{\theta}(p) = \dot{\theta}(0)$. Hence, again from the equations for r and $\mathbf{r}$, we find $\dot{\mathbf{r}}(p) = \dot{\mathbf{r}}(0)$. The uniqueness statement in the Picard–Lindelöf theorem, applied to the system (K′) of two equations of first order, tells us that $\mathbf{r}(t+p) = \mathbf{r}(t)$ for all $t \in \mathbb{R}$, see Exercise 1.1.

Let $p' \in \mathbb{R}^+$ be any other period of $\mathbf{r}$. Since $\dot{\theta} > 0$, this implies $\theta(p') = \theta(0) + 2n\pi$ for some $n \in \mathbb{N}$, and the period p, corresponding to $n = 1$, must be the smallest of these positive periods. □

Theorem 3.7 (Kepler's third law) *For elliptic solutions of the Kepler problem, the relation between semi-major axis a and minimal period p is given by*

$$\boxed{\frac{p^2}{a^3} = \frac{4\pi^2}{\mu}.} \tag{3.9}$$

Proof The area A enclosed by the ellipse

$$\left\{(x,y) \in \mathbb{R}^2 : \frac{x^2}{a^2} + \frac{y^2}{b^2} = 1\right\}$$

with semi-major axis a and semi-minor axis $b = a\sqrt{1-e^2}$ can be computed as twice the area under the graph of the function

$$y = b\sqrt{1 - \frac{x^2}{a^2}}, \quad -a \le x \le a,$$

that is,

$$A = 2b \int_{-a}^{a} \sqrt{1 - \frac{x^2}{a^2}}\, \mathrm{d}x.$$

With the substitution $x = a\cos\varphi$, where φ ranges from π to 0, and $\mathrm{d}x = -a\sin\varphi\,\mathrm{d}\varphi$, this becomes

$$A = 2ab \int_{0}^{\pi} \sin^2\varphi\,\mathrm{d}\varphi = \pi ab.$$

On the other hand, Proposition 1.5 tells us that

$$A = \frac{1}{2}\int_{0}^{p} r^2\dot{\theta}\,\mathrm{d}t = \frac{1}{2}pc.$$

Comparing these two expressions for A, we find $pc = 2\pi a^2\sqrt{1-e^2}$. By squaring this equation, and with the help of (3.4), we easily deduce (3.9). □

Notes and references

Kepler's first law was originally proved by Newton (1687). In Propositions 11 to 13 of Book 1 of his *Principia*, he solves what we would now call the inverse problem: given that a body moves in a central force field along a conic section with the force centre at one of the foci, show that the force law is $F \propto r^{-2}$. The direct problem – describe the trajectories for the given force law $F \propto r^2$ – is treated somewhat laconically in Corollary 1 to Proposition 13.

The proof of Kepler's first law in Section 3.1 is essentially due to Laplace, except that he did not have vector notation at his disposal. A good source for this proof is (Somerville, 1831), which is a "greatly adapted and simplified translation of Laplace's multi-volume *Mécanique Céleste*," written at the instigation of the *Society for the Diffusion of Useful Knowledge* "to give popular audiences an introduction to astronomical methods and theories." Laplace's proof of Kepler's first law is contained in paragraphs 365–371. Equation (3.1) in Section 3.1 is equation (91) on page 187 of Somerville's book; equation (3.3) appears towards the end of paragraph 371. For an appreciation of Somerville's achievement in making Laplace's work more accessible, see the fascinating historical study (Secord, 2014).

Kepler's third law is Proposition 15 in Book 1 of (Newton, 1687). The converse of Kepler's third law for circular motions (if $p \propto a^{3/2}$, then the attractive force is proportional to r^{-2}) is stated in Corollary 6 to Proposition 4 in Book 1.

In a scholium Newton adds: "The case of corol. 6 holds for the heavenly bodies (as our compatriots Wren, Hooke and Halley have also found out independently)." One may well, as John Aubrey did, sympathise with Hooke, who felt that this statement did not properly acknowledge his role in the discovery of the inverse square law. The letter alluded to by Aubrey is Hooke's letter to Newton of 6 January 1680. For further information on the correspondence between Hooke and Newton and rival claims to priority concerning the inverse square law see (Koyré, 1952).

Exercises

3.1 Show that the scalar product $\langle \mathbf{a} \times \mathbf{b}, \mathbf{c} \rangle$ equals the oriented volume of the parallelepiped spanned by the vectors $\mathbf{a}, \mathbf{b}, \mathbf{c} \in \mathbb{R}^3$. Use this to prove identity (3.2).

3.2 (a) Let $\mathbf{r}$ be a solution of the Kepler problem (K). By the conservation of angular momentum this is a planar curve, so we may write $\mathbf{r} = r(\cos\theta, \sin\theta)$. Show that

$$r^2\ddot{r} - \frac{c^2}{r} = -\mu, \tag{3.10}$$

where $c = r^2\dot{\theta}$.

(b) Show that, conversely, a C^2-curve $t \mapsto r(t)\,(\cos\theta(t), \sin\theta(t))$ with $c := r^2\dot{\theta}$ constant is a solution of (K) if it satisfies (3.10).

(c) Consider a C^2-curve $t \mapsto r(t)\,(\cos\theta(t), \sin\theta(t))$ with $c := r^2\dot{\theta}$ a positive constant. Then r may be regarded as a function of θ. Show that

$$r(\theta) := \frac{c^2/\mu}{1 + e\cos\theta}$$

defines a solution of (K); see Exercise 1.6. In particular, this proves the converse to Kepler's first law: if motions in a central force field obey Kepler's first law, then the force must be proportional to r^{-2}.

3.3 Let $\mathbf{r}$ be a solution of (K) with $c = r^2\dot{\theta} \neq 0$. Set $\rho = 1/r$.

(a) Show that $\dot{r} = -c\rho'$ and $\ddot{r} = -c^2\rho^2\rho''$, where the prime denotes derivative with respect to θ.

(b) Conclude that ρ satisfies the differential equation

$$\rho'' + \rho = \frac{\mu}{c^2}.$$

(c) Deduce, once again, Kepler's first law by finding the solution to this simple inhomogeneous linear differential equation.

3.4 In this exercise we shall see yet another (and very elegant) proof of Kepler's first law, this one due to (Lagrange, 1785, §14). Let $\mathbf{r} = (x, y)$ be a solution of (K). Written in components, this means

$$\ddot{x} = -\mu \frac{x}{r^3} \quad \text{and} \quad \ddot{y} = -\mu \frac{y}{r^3}.$$

Moreover, equation (3.10) can be rewritten as

$$\ddot{r} = -\mu \frac{r}{r^3} + \frac{c^2}{r^3}.$$

In other words, r is a solution of the inhomogeneous linear differential equation

$$\ddot{\xi} = -\mu \frac{\xi}{r^3(t)} + \frac{c^2}{r^3(t)},$$

whereas x and y are solutions of the corresponding homogeneous linear differential equation

$$\ddot{\xi} = -\mu \frac{\xi}{r^3(t)}.$$

Observe that $\xi = c^2/\mu$ is likewise a solution of the inhomogeneous equation.

Let us now assume $c \neq 0$, even though the following argument can be adapted to the case $c = 0$.

(a) Use standard results from the theory of linear differential equations to argue that there must be real constants α, β such that

$$r(t) = \frac{c^2}{\mu} + \alpha x(t) + \beta y(t).$$

Why is the assumption $c \neq 0$ essential here?
We conclude that, in euclidean 3-space with cartesian coordinates (x, y, r), the solution $t \mapsto (x(t), y(t), r(t))$ lies on the conic section given by intersecting the cone

$$\{(x, y, r) \in \mathbb{R}^3 : r^2 = x^2 + y^2\}$$

with the plane

$$\{(x, y, r) \in \mathbb{R}^3 : r = c^2/\mu + \alpha x + \beta y\}.$$

(b) The solution curve $t \mapsto (x(t), y(t))$ of (K) then satisfies the equation

$$\sqrt{x^2 + y^2} = \frac{c^2}{\mu} + \alpha x + \beta y. \tag{3.11}$$

Set $e := \sqrt{\alpha^2 + \beta^2}$, and let F be the origin $(0, 0)$ in the xy-plane. If $e = 0$, equation (3.11) clearly describes a circle centred at F.
If $e \neq 0$, let ℓ be the line

$$\ell := \{(x, y) \in \mathbb{R}^2 : c^2/\mu + \alpha x + \beta y = 0\}.$$

Convince yourself that equation (3.11) can then be read as

$$|PF| = e \operatorname{dist}(P, \ell),$$

which by Corollary 2.16 describes a conic section with one focus at the origin F.

3.5 Let $\mathbf{r}$ be a solution of (K) with $c \neq 0$. Show that in the limit $t \to \pm\infty$, the speed v goes to zero for $h = 0$, and v reaches a positive limit for $h > 0$. What happens for $c = 0$?

3.6 (a) Let $G \neq \{0\}$ be a subgroup of $(\mathbb{R}, +)$. Set $\alpha := \inf(G \cap \mathbb{R}^+)$. Show the following.

(1) If $\alpha = 0$, then G is dense in $\mathbb{R}$.
(2) If $\alpha > 0$, then $G = \alpha\mathbb{Z}$.

(b) Let $f : \mathbb{R} \to \mathbb{R}^n$ be any function. A number $p \in \mathbb{R} \setminus \{0\}$ is called a **period** of f if $f(t + p) = f(t)$ for all $t \in \mathbb{R}$. Set

$$\mathcal{P} := \{p \in \mathbb{R} : p \text{ is a period of } f\} \cup \{0\}.$$

Show the following.

(1) $\mathcal{P}$ is a subgroup of $(\mathbb{R}, +)$.
(2) If f is continuous and not constant, then $\mathcal{P} = \alpha\mathbb{Z}$ for some $\alpha > 0$.

(c) Let $\chi : \mathbb{R} \to \mathbb{R}$ be the characteristic function of $\mathbb{Q}$, that is, $\chi(x) = 1$ for $x \in \mathbb{Q}$ and $\chi(x) = 0$ for $x \in \mathbb{R} \setminus \mathbb{Q}$. Determine $\mathcal{P}$ for this function.

3.7 Let $t \mapsto \mathbf{r}(t)$, $t \in \mathbb{R}$, be a solution of the Kepler problem with energy $h < 0$. Let $\mathcal{E}$ be the curve traced out by the body in the plane through $\mathbf{0}$ and orthogonal to the angular momentum $\mathbf{c}$. We wish to give an alternative proof that this solution curve $\mathcal{E}$ is an ellipse, following ideas of (van Haandel and Heckman, 2009). To this end, let C be the circle of radius $-\mu/h$ about $\mathbf{0}$ in the orbital plane. For total energy h, this circle corresponds to points of kinetic energy zero, so $\mathcal{E}$ has to lie in the interior of C. Now take a look at Figure 3.2.

Let $\mathbf{s}$ be the projection of the point $\mathbf{r}$ from $\mathbf{0}$ to C, i.e. $\mathbf{s} := -\mu\mathbf{r}/hr$. Let $\mathbf{A}$ be the point obtained by reflecting the point $\mathbf{s}$ in the tangent $\mathcal{T}$ to $\mathcal{E}$ at $\mathbf{r}$. Show the following.

(a) With $\mathbf{n} := \mathbf{v} \times \mathbf{c}$ we have $\mathbf{A} = \mathbf{s} - 2(\langle \mathbf{s} - \mathbf{r}, \mathbf{n}\rangle \cdot \mathbf{n})/n^2$.

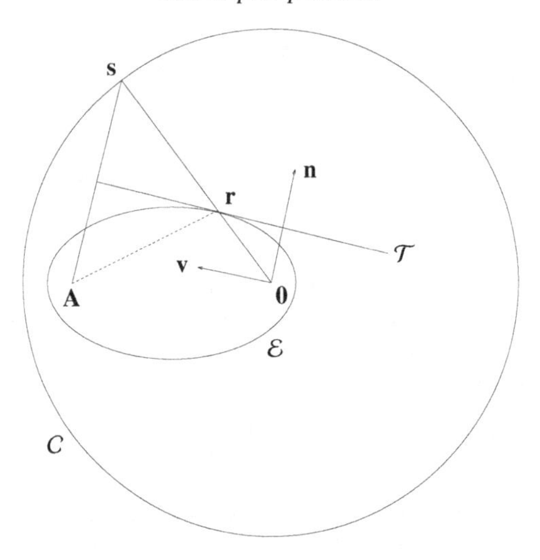

Figure 3.2 An alternative proof of Kepler's first law.

(b) $\langle \mathbf{s} - \mathbf{r}, \mathbf{n} \rangle = -(1 + \mu/hr)c^2$.
(c) $n^2 = c^2v^2 = 2c^2(h + \mu/r)$.
(d) $\mathbf{A} = -\mu\mathbf{r}/hr + \mathbf{n}/h$.
(e) $\dot{\mathbf{A}} = \mathbf{0}$, i.e. the point $\mathbf{A}$ is independent of t.
(f) $\mathcal{E}$ is the ellipse with foci $\mathbf{0}$ and $\mathbf{A}$, and semi-major axis $a = -\mu/2h$.

3.8 In this exercise we consider light rays in the plane and their reflection in a mirror described by a smooth curve $\mathcal{M}$.

(a) Let $\mathcal{M}$ be a straight line in the plane and A, B two points in the plane on the same side of $\mathcal{M}$. Show that the shortest path from A to B via a point $P \in \mathcal{M}$ is the one made up of two line segments AP and PB that make the same angle with $\mathcal{M}$.
Hint: What is the shortest path in the plane from A to the point B' obtained by reflecting B in $\mathcal{M}$?

(b) Let A, B be two points in the plane, and consider the function $P \mapsto f(P) := |AP| + |PB|$, $P \in \mathbb{R}^2 \setminus \{A, B\}$. Use Exercise 1.8 to show that grad $f(P)$ equals the sum of the unit vector in the direction from A to P and the unit vector in the direction from B to P.

(c) Deduce with the Lagrange multiplier theorem that the critical points of f subject to the constraint $P \in \mathcal{M}$, where $\mathcal{M}$ is any smooth curve in the plane, are the ones where AP and PB make the same angle with the tangent to $\mathcal{M}$ at P.

3.9 At the heart of Exercise 3.7 is the *reflection property* of ellipses. Let $\mathcal{E} \subset \mathbb{R}^2$ be an ellipse with foci F_1, F_2. Let $P \in \mathcal{E}$ be a point on the ellipse, and $\mathcal{T}$ the tangent line to $\mathcal{E}$ in P. Show that the angle between PF_1 and $\mathcal{T}$ equals that between PF_2 and $\mathcal{T}$. By the law of reflection from the preceding exercise, this means that a light ray emanating from F_1 will be reflected at the ellipse towards F_2.

Hint: Convince yourself that the path from F_1 to F_2 via P is the shortest path between the two foci via a point on $\mathcal{T}$. Alternatively, the preceding exercise can be used more directly by observing that $\mathcal{E}$ is a level set of the function $P \mapsto |F_1P| + |F_2P|$.

3.10 Use the analogue of Exercise 3.8 (b) and the gardener's construction of parabolas and hyperbolas to derive a reflection property of these curves.

4

The dynamics of the Kepler problem

Newtonian, *adj.* Pertaining to a philosophy of the universe, invented by Newton, who discovered that an apple will fall to the ground, but was unable to say why. His successors and disciples have advanced so far as to be able to say when.

Ambrose Bierce, *The Devil's Dictionary*

In Chapter 3 we solved the Kepler problem to the extent that we were able to determine the geometric trajectory of a body moving in a gravitational field. However, as yet we do not know the position of the body as a function of time. In this chapter we show that the trajectory can be parametrised by a suitable angular variable, the so-called eccentric anomaly, which can be given at least implicitly as a function of time. In the elliptic and hyperbolic case, this so-called *Kepler equation* is transcendental and cannot be solved explicitly in terms of elementary functions.

In Section 4.1 we derive Kepler's equation in the elliptic case. The analogous hyperbolic case is relegated to the exercises. In Section 4.2 I present Newton's original solution of Kepler's equation (in the elliptic case) by means of a geometric construction involving a cycloid.

In the parabolic case, the relevant equation is cubic. I take this as an excuse to make a small detour, in Section 4.3, to the algebra of cubic equations and their explicit solution by radicals. This relies on an elementary decomposition of the cube, very much analogous to the solution of quadratic equations by quadratic extension.

4.1 Anomalies and Kepler's equation

In this section we concentrate on the elliptic case. Thus, let $\mathbf{r}\colon \mathbb{R} \to \mathbb{R}^3 \setminus \{\mathbf{0}\}$ be a solution of the Kepler problem with $h < 0$ and $c \neq 0$.[1] As in Section 3.3, we may take $\mathbf{r}$ to describe an ellipse $\mathcal{E}$ in the xy-plane, traversed positively. Moreover, we can choose our coordinates such that the eccentricity vector $\mathbf{e}$ points in the positive x-direction. Then the angle ω, as defined in Figure 3.1, is zero, and the true anomaly f describes the angle between the positive x-axis and $\mathbf{r}$. Hence, writing $\mathbf{r}$ as a function of f – which is permissible since $\dot{f} > 0$ – we have

$$\mathbf{r}(f) = \frac{a(1-e^2)}{1+e\cos f}(\cos f, \sin f), \quad f \in \mathbb{R}.$$

Now draw a circle C with centre $Z = (-ea, 0)$, which equals the centre of $\mathcal{E}$, and radius a. To any point $P \in \mathcal{E}$ there corresponds a unique point $P' \in C$ as shown in Figure 4.1.

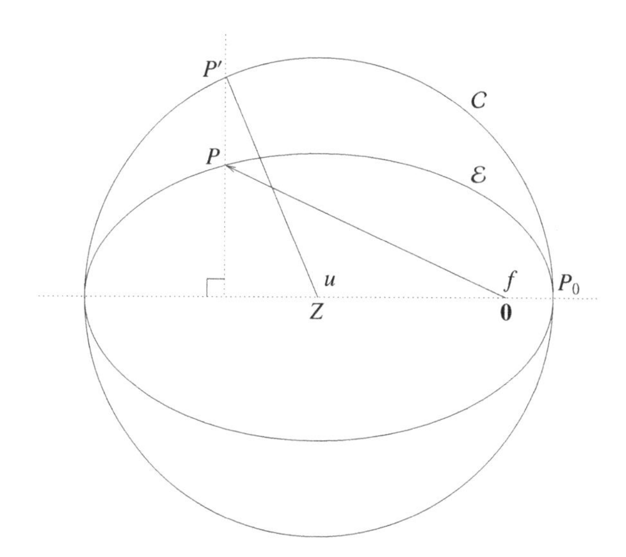

Figure 4.1 The true anomaly f and the eccentric anomaly u.

Definition 4.1 The **eccentric anomaly** of $P \in \mathcal{E}$ is the angle[2] $u := \angle \mathbf{0}ZP'$. The point $P_0 \in \mathcal{E}$ corresponding to $u \in 2\pi\mathbb{Z}$ (equivalently, $f \in 2\pi\mathbb{Z}$), where $r = r_{\min}$ in the notation of Section 2.1, is called the **pericentre** of $\mathcal{E}$.

[1] We disregard the trivial circular case $e = 0$. The case $h < 0$ and $c = 0$ is discussed in Exercise 4.7.

[2] The notation u for the eccentric anomaly is due to historical conventions.

The notion of pericentre is defined analogously for hyperbolas and parabolas. Thinking dynamically, if a curve $\mathbf{r}$ traces out a conic section with one focus at $\mathbf{0}$, a **pericentre passage** is any moment t_0 where $\mathbf{r}(t_0)$ lies at the pericentre, i.e. where $r(t_0) = r_{\min}$.

Lemma 4.2 *In a cartesian coordinate system with origin at one of the foci of the ellipse $\mathcal{E}$, and the positive x-direction determined by the eccentricity vector $\mathbf{e}$, the point $P \in \mathcal{E}$ with eccentric anomaly u has the coordinates*

$$(a\cos u - ea, b\sin u) = a\,(\cos u - e, \sqrt{1-e^2}\,\sin u).$$

Proof The corresponding point $P' \in C$ has the coordinates

$$(a\cos u - ea, a\sin u).$$

The point P has the same x-coordinate as P'. The y-coordinates of P and P' are related by $y(P) = y(P') \cdot b/a$, as is clear from comparing the cartesian descriptions of C and $\mathcal{E}$: the former is given by $(x+ea)^2/a^2 + (y/a)^2 = 1$, the latter by $(x+ea)^2/a^2 + (y/b)^2 = 1$. □

Observe that the positive traversal of $\mathcal{E}$ corresponds to increasing u.

Proposition 4.3 *The planar curve $\mathbf{r}\colon \mathbb{R} \to \mathbb{R}^2 \setminus \{\mathbf{0}\}$ is a solution of the Kepler problem* (K) *with*

$$\begin{cases} c \neq 0 \\ h < 0 \\ \text{semi-major axis } a \\ \mathbf{e} \text{ pointing in the positive } x\text{-direction} \\ \text{pericentre passage for } t = t_0 \end{cases}$$

if and only if

$$\mathbf{r}(t) = a\,(\cos u(t) - e, \sqrt{1-e^2}\,\sin u(t)),$$

where u is the unique solution of the **Kepler equation**

$$\boxed{u(t) - e\sin u(t) = \frac{2\pi}{p}(t - t_0).} \tag{4.1}$$

Here p is the minimal period determined by Kepler's third law.

Definition 4.4 The term $(2\pi/p)(t - t_0)$ is called the **mean anomaly**.

Proof of Proposition 4.3 Let $\mathbf{r}$ be a solution of (K) with the given orbit data. By Lemma 4.2, $\mathbf{r}$ can be written as

$$\mathbf{r}(t) = a\,(\cos u(t) - e, \sqrt{1-e^2}\,\sin u(t)),$$

where $u\colon \mathbb{R} \to \mathbb{R}$ is a C^2-function, cf. Remark 1.4. We want to show that u solves the Kepler equation. We compute

$$\dot{\mathbf{r}} = a\,(-\dot{u}\sin u,\ \sqrt{1-e^2}\,\dot{u}\cos u).$$

If we regard $\mathbf{r}$ and $\dot{\mathbf{r}}$ as vectors in $\mathbb{R}^2 \equiv \mathbb{R}^2 \times \{0\} \subset \mathbb{R}^3$, then $\mathbf{c} = \mathbf{r} \times \dot{\mathbf{r}} = (0, 0, c)$ with

$$\begin{aligned} c &= a^2\sqrt{1-e^2}\,\dot{u}\,((\cos u - e)\cos u + \sin^2 u) \\ &= a^2\sqrt{1-e^2}\,\dot{u}\,(1 - e\cos u). \end{aligned}$$

From (3.4) we have $c = \sqrt{\mu a(1-e^2)}$, hence

$$\dot{u} = \frac{\sqrt{\mu}}{a^{3/2}} \cdot \frac{1}{1 - e\cos u}. \tag{4.2}$$

In particular, we have

$$\dot{u} \geq \frac{\sqrt{\mu}}{a^{3/2}} \cdot \frac{1}{1+e} > 0,$$

which implies that $u\colon \mathbb{R} \to \mathbb{R}$ is a diffeomorphism.[3] So there is a unique $t_0 \in \mathbb{R}$ such that $u(t_0) = 0$, corresponding to the pericentre $\mathbf{r}(t_0) = (a(1-e), 0)$.

Rewrite (4.2) as

$$\dot{u}(1 - e\cos u) = \frac{\sqrt{\mu}}{a^{3/2}}.$$

Integrating this equation gives

$$u(t) - e\sin u(t) = \frac{\sqrt{\mu}}{a^{3/2}}(t - t_0).$$

Setting $t = t_0 + p$ we obtain the equation

$$2\pi - e\sin 2\pi = \frac{\sqrt{\mu}}{a^{3/2}}p,$$

from which we recover Kepler's third law $p^2/a^3 = 4\pi^2/\mu$. The Kepler equation follows. Observe that

$$\frac{\mathrm{d}}{\mathrm{d}u}(u - e\sin u) = 1 - e\cos u > 0,$$

so, by the inverse function theorem, the Kepler equation does indeed determine a unique (and smooth) function u. A direct computation shows that the solution u of the Kepler equation gives rise to a solution $\mathbf{r}$ of the Kepler problem, see Exercise 4.1. □

For the corresponding statement in the hyperbolic case, see Exercise 4.3. The parabolic case will be treated in Section 4.3.

[3] This gives an alternative proof of Corollary 3.3 in the elliptic case.

4.2 Solution of Kepler's equation by the cycloid

For a body moving along an elliptic Keplerian orbit, we now describe a geometric construction that allows one to locate the position of the body at any given moment in time. For $e = 0$ the Kepler equation (or Kepler's second law) tells us that the body moves with constant angular velocity along its circular orbit. So from now on we assume $e \neq 0$.

Take a look at Figure 4.2. We consider an ellipse $\mathcal{E}$ with semi-major axis a and eccentricity e, centred at the point $Z = (0, a/e)$, and with major axis along the y-axis. Also, we draw a circle C of radius a about Z, as in Figure 4.1, and a further circle C' of radius a/e. So the circle C' touches the x-axis at the origin.[4] The orientation of the x-axis has been chosen to the left for practical purposes.

Now roll the disc bounded by C' to the left along the x-axis. Then the point $P_0 = (0, a/e - a)$, the pericentre of $\mathcal{E}$ relative to the focus $F = (0, a/e - ea)$, describes the curve shown in Figure 4.2, which is called a **curtate cycloid**. (A point on C' would describe a curve simply known as the **cycloid**.)

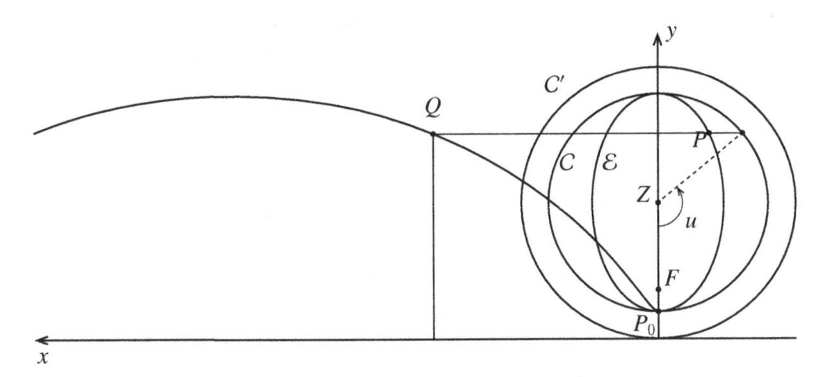

Figure 4.2 Newton's solution of the Kepler equation using a curtate cycloid.

Lemma 4.5 *The curtate cycloid is parametrised by*

$$u \longmapsto (x, y) = \left(\frac{a}{e}\,u - a\sin u, \frac{a}{e} - a\cos u\right),$$

where $u \in \mathbb{R}$ is the angle through which the rolling disc has rotated.

Proof When the disc of radius a/e has rotated through an angle u, its centre lies at the point $(ua/e, a/e)$. Given our choice of coordinates, the position of the point on C tracing out the cycloid, relative to the centre of the disc, is described by the vector $-a(\sin u, \cos u)$. □

[4] By the computation in the proof of Proposition 2.5, the x-axis is the directrix of $\mathcal{E}$.

In this parametrisation, the point P_0 corresponds to $u = 0$; the point Q, to the angle u shown in Figure 4.2.

We now think of F as the focus of $\mathcal{E}$ containing the force centre, so that the angle u is the eccentric anomaly of the point P. Our aim is to find that point as a function of time. As before, we take t_0 as a time of a pericentre passage, i.e. $P(t_0) = P_0$.

Let $Q = Q(t)$ be the point on the curtate cycloid with x-coordinate

$$2\pi \frac{a}{e} \cdot \frac{t - t_0}{p}. \tag{4.3}$$

In other words, we take the distance along the x-axis whose ratio to the circumference of C' equals the ratio between the time $t - t_0$ elapsed since a pericentre passage and the period p of the elliptic motion. Observe that

$$\frac{\mathrm{d}x}{\mathrm{d}u} = \frac{a}{e} - a\cos u > 0,$$

so there is indeed a unique such point Q.

The horizontal line through Q intersects $\mathcal{E}$ at just a single point $P = P(t)$ if $t - t_0$ is an integer multiple of $p/2$ (corresponding to u being an integer multiple of π), and at two points otherwise. We take the intersection point to the right or to the left of the y-axis if $(t - t_0)/p$ minus its integer part lies in the open interval $(0, 1/2)$ or $(1/2, 1)$, respectively, i.e. we think of the ellipse as being traversed counter-clockwise.

Proposition 4.6 *In the Keplerian motion along $\mathcal{E}$, the point $P = P(t)$ is the position of the body at time t.*

Proof Our construction produces a point $P(t) \in \mathcal{E}$ with eccentric anomaly $u(t)$. Comparing our choice (4.3) of x-coordinate for Q with the parametrisation of the curtate cycloid in Lemma 4.2, we find

$$\frac{a}{e}\, u(t) - a \sin u(t) = 2\pi \frac{a}{e} \cdot \frac{t - t_0}{p},$$

which after dividing by a/e is the Kepler equation. □

4.3 The parabolic case: cubic equations

We now want to investigate solutions $\mathbf{r}\colon \mathbb{R} \to \mathbb{R}^2 \setminus \{\mathbf{0}\}$ of the Kepler problem with $h = 0$ and $c \neq 0$, i.e. the parabolic case. Choose cartesian coordinates in $\mathbb{R}^2$ such that the unit vector $\mathbf{e}$ points in the positive x-direction. Then equation (3.3) becomes

$$\sqrt{x^2+y^2} = -x + \frac{c^2}{\mu}.$$

Since $\sqrt{x^2+y^2} \geq |x| \geq 0$ (and $c^2/\mu > 0$), the equation

$$\sqrt{x^2+y^2} = x - \frac{c^2}{\mu}$$

does not have any solutions, so the previous equation is equivalent to the one obtained by squaring:

$$y^2 = \frac{c^4}{\mu^2} - \frac{2c^2}{\mu}x.$$

This means that the trajectory can be parametrised by

$$u \longmapsto \left(\frac{1}{2}\left(\frac{c^2}{\mu} - u^2\right), \frac{c}{\sqrt{\mu}}u\right), \quad u \in \mathbb{R}.$$

We call u the **eccentric anomaly** in the parabolic case. This parametrisation has been chosen such that positive traversal corresponds to increasing u. The following is the parabolic analogue of Proposition 4.3.

Proposition 4.7 *The planar curve* $\mathbf{r}\colon \mathbb{R} \to \mathbb{R}^2 \setminus \{\mathbf{0}\}$ *is a solution of the Kepler problem* (K) *with*

$$\begin{cases} c \neq 0 \\ h = 0 \\ \mathbf{e} \text{ pointing in the positive } x\text{-direction} \\ \text{pericentre passage for } t = t_0 \end{cases}$$

if and only if

$$\mathbf{r}(t) = \left(\frac{1}{2}\left(\frac{c^2}{\mu} - u^2(t)\right), \frac{c}{\sqrt{\mu}}\, u(t)\right),$$

where $u(t)$ is the unique real solution of the cubic equation

$$\boxed{\frac{1}{6}u^3(t) + \frac{c^2}{2\mu}\, u(t) = \sqrt{\mu}\,(t - t_0).}$$

Proof Let $\mathbf{r}$ be a solution of (K) with the given orbit data. According to the discussion preceding the proposition, $\mathbf{r}$ can be written as

$$\mathbf{r}(t) = \left(\frac{1}{2}\left(\frac{c^2}{\mu} - u^2(t)\right), \frac{c}{\sqrt{\mu}}\, u(t)\right),$$

where $u\colon \mathbb{R} \to \mathbb{R}$ is a C^2-function. We want to show that u solves the cubic equation in the proposition. We compute

$$\dot{\mathbf{r}} = (-u\dot{u}, c\dot{u}/\sqrt{\mu}).$$

Again we regard $\mathbf{r}$ and $\dot{\mathbf{r}}$ as vectors in $\mathbb{R}^2 \equiv \mathbb{R}^2 \times \{0\} \subset \mathbb{R}^3$. Then $\mathbf{r} \times \dot{\mathbf{r}} = (0, 0, c)$ with

$$\begin{aligned} c &= \frac{c}{2\sqrt{\mu}}\,\dot{u}\left(\frac{c^2}{\mu} - u^2\right) + \frac{c}{\sqrt{\mu}}\,u^2\dot{u} \\ &= \frac{c}{2\sqrt{\mu}}\,\dot{u}\left(u^2 + \frac{c^2}{\mu}\right), \end{aligned}$$

which is equivalent to

$$\frac{1}{2}\dot{u}u^2 + \frac{c^2}{2\mu}\,\dot{u} = \sqrt{\mu}.$$

We see that $\dot{u} > 0$ (which is consistent with positive traversal), and from Corollary 3.3 we know that $u\colon \mathbb{R} \to \mathbb{R}$ is actually a diffeomorphism. So there is a unique $t_0 \in \mathbb{R}$ with $u(t_0) = 0$. Thus, by integration we obtain

$$\frac{1}{6}u^3 + \frac{c^2}{2\mu}\,u = \sqrt{\mu}\,(t - t_0).$$

For the converse statement in the proposition see Exercise 4.4. □

The following proposition shows that the real solution of the cubic equation in the preceding proposition can be written down explicitly in terms of radicals.

Proposition 4.8 *Let $p > 0$ and q be real parameters. The cubic equation*

$$x^3 + 3px + q = 0$$

has the unique real solution

$$x = \frac{p}{a} - a,$$

where

$$a = \sqrt[3]{\frac{q}{2} + \sqrt{\frac{q^2}{4} + p^3}}.$$

Proof Figure 4.3 illustrates the identity

$$x^3 + 3aAx + a^3 = A^3$$

for $A = x + a$. This geometric interpretation presupposes that x and a are positive real numbers, but the identity is true regardless of signs, or even for

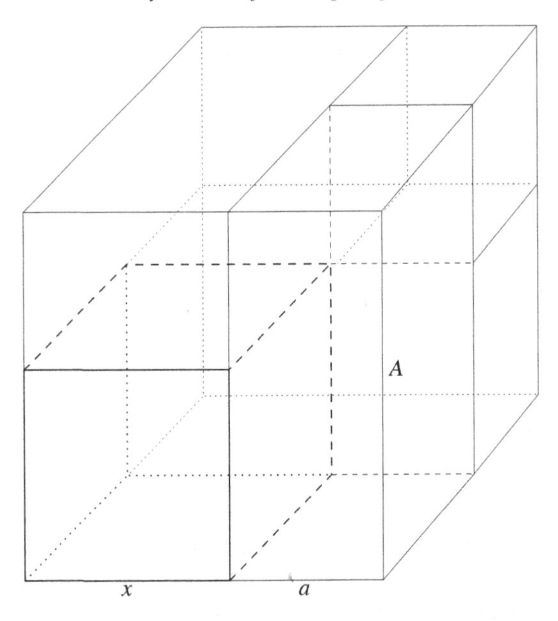

Figure 4.3 Decomposition of the cube.

complex numbers. So the cubic equation $x^3 + 3px + q = 0$ can be rewritten as the following system of equations:

$$\left.\begin{array}{rcl} x + a & = & A \\ aA & = & p \\ a^3 - A^3 & = & q. \end{array}\right\}$$

Multiplying the third equation by a^3 and substituting $aA = p$ yields

$$a^6 - qa^3 - p^3 = 0.$$

This is a quadratic equation in a^3, which has the solutions

$$a^3 = \frac{q}{2} \pm \sqrt{\frac{q^2}{4} + p^3}.$$

Since $p > 0$, the positive real solution a of this last equation is as stated in the proposition, and the claimed expression of a real solution x follows. Since, for $p > 0$, the cubic polynomial is strictly monotone in x, this is the unique real solution. □

Remark 4.9 Computing over the complex numbers (with arbitrary $p, q \in \mathbb{C}$), one finds six complex solutions a (counted with multiplicities) of the last equation in the above proof. Pairwise they yield the same solution x, so there are indeed three solutions of the cubic equation over $\mathbb{C}$, see Exercise 4.5.

Notes and references

Section 4.2 is based on Proposition 31 in Book 1 of (Newton, 1687). In a scholium to that proposition, Newton notes that "the description of this curve [the curtate cycloid] is difficult; hence it is preferable to use a solution that is approximately true." He then goes on to describe what is now known as Newton's method for finding the zeros of a function by sequential approximation. This technique is explained in most introductory texts on analysis; in the context of Kepler's equation the method is discussed in (Danby, 1992) or (Ortega and Ureña, 2010). Hundreds of papers have been written on analytical and numerical solutions of Kepler's equation. Danby's book contains a number of solutions, and it discusses the convergence rates of various numerical approaches. Regarding analytical solutions, both of the references cited discuss a series expansion in terms of Bessel functions, see also (Pollard, 1966).

The formula for solving cubic equations (Proposition 4.8) is known as Cardano's formula, after the Italian polymath Gerolamo Cardano (1501–1576), who published this solution in his 1545 book *Artis Magnae*. The essential ideas were in fact known earlier to Scipione del Ferro (1465–1526) and Niccolò Tartaglia (1500–1557), from whom Cardano had learned the solution. I have included a proof because I feel that most algebra texts give insufficient emphasis to the geometric idea behind this formula.

One important aspect of the Kepler problem not treated in this text is the question of how to determine an orbit from just a few observations. For instance, can one determine the Keplerian orbit passing through two given positions, if the time taken for the transfer between those two positions is known? This is not only of historical significance, but also essential in the design of space missions. A good introduction to this topic is given in Chapters 6 and 7 of (Danby, 1992). The specific question stated above leads to a result known as Lambert's theorem, see (Danby, 1992, Section 6.11) and (Pollard, 1966, Section 1.11). This theorem states that the travel time between two points along an elliptic trajectory is a function only of the linear distance between the two points, the energy (or, by equation (3.8), the semi-major axis), and the *sum* of the distances of the two points from the force centre. Surprisingly, the individual distances from the force centre need not be known. For a lucid explanation of the significance of Lambert's theorem, as well as further material on the Kepler problem, see (Albouy, 2002).

Exercises

4.1 In this exercise you will be asked to verify that solutions of the Kepler equation do indeed give rise to a solution of the Kepler problem in the elliptic case, see Proposition 4.3.

(a) Let $a, \mu > 0$, $0 \leq e < 1$ and $t_0 \in \mathbb{R}$ be given. Let $u\colon \mathbb{R} \to \mathbb{R}$ be the diffeomorphism determined by the Kepler equation (4.1), with p as in (3.9). Show that

(1) $\dot{u} = \dfrac{2\pi}{p(1 - e\cos u)}$,

(2) $\ddot{u} = -\dfrac{4\pi^2}{p^2} \cdot \dfrac{e \sin u}{(1 - e\cos u)^3}$.

(b) Set

$$\mathbf{r}(t) := a\,(\cos u(t) - e,\ \sqrt{1 - e^2}\,\sin u(t)).$$

Show that $r = a(1 - e\cos u)$ and verify that $\mathbf{r}$ is a solution of the Kepler problem (K).

4.2 Let $\mathbf{r}$ be a solution of the Kepler problem with $c \neq 0$ and $h < 0$ (elliptic case). Starting from the equation

$$a(1 - e\cos u) = r = \frac{a(1 - e^2)}{1 + e\cos f},$$

see the foregoing exercise, show that the true anomaly f and the eccentric anomaly u are related by

$$\tan\frac{f}{2} = \sqrt{\frac{1 + e}{1 - e}} \cdot \tan\frac{u}{2}.$$

4.3 In this exercise we prove the analogue of Proposition 4.3 in the hyperbolic case. Let $\mathbf{r}\colon \mathbb{R} \to \mathbb{R}^2 \setminus \{\mathbf{0}\}$ be a solution of the Kepler problem (K) with

$$\begin{cases} c \neq 0 \\ h > 0 \\ \text{real semi-axis } a \\ \mathbf{e} \text{ pointing in the positive } x\text{-direction} \\ \text{pericentre passage for } t = t_0. \end{cases}$$

Set $p = 2\pi a^{3/2}/\sqrt{\mu}$ as in the elliptic case. (Beware, though, that in the hyperbolic case we cannot speak of a period.) According to Exercise 2.7 we can write $\mathbf{r}$ as

$$\mathbf{r}(t) = a(e - \cosh u(t),\ \sqrt{e^2 - 1}\,\sinh u(t)).$$

We call u the **eccentric anomaly** in the hyperbolic case.

(a) Compute r.

(b) Show that u satisfies the **Kepler equation**

$$\boxed{e \sinh u(t) - u(t) = \frac{2\pi}{p}(t - t_0).}$$

Conversely, explain why this equation determines a unique smooth function u of t, and show that this yields a solution of the Kepler problem.

4.4 Explain why the cubic equation in Proposition 4.7 determines a *smooth* function u of t. Then define $\mathbf{r}(t)$ as in that proposition, and verify that this yields a solution of the Kepler problem.

4.5 (a) Show that any cubic equation

$$y^3 + ay^2 + by + c = 0$$

over $\mathbb{C}$ can be reduced to the form $x^3 + 3px + q = 0$ by a suitable linear substitution $y = x + d$.

(b) Perform the argument in the proof of Proposition 4.8 over $\mathbb{C}$ and verify that the six solutions for a reduce to three solutions x. Write down these three solutions explicitly.

4.6 Show that for solutions $\mathbf{r}$ of the Kepler problem with $c \neq 0$, the eccentric anomaly u satisfies the equation

$$u(t) = \sqrt{\mu} \int_{t_0}^{t} \frac{\mathrm{d}\tau}{r(\tau)} \quad \text{for } h = 0, \text{ resp. } \quad u(t) = \sqrt{\frac{\mu}{a}} \int_{t_0}^{t} \frac{\mathrm{d}\tau}{r(\tau)} \quad \text{for } h \neq 0.$$

Hint: In the elliptic case, look at (4.2). Derive a similar formula in the hyperbolic case.

4.7 From the proof of Proposition 4.3 and part (b) of Exercise 4.1 we see that any solution of the Kepler problem with $h < 0$, $c \neq 0$ (and hence $e < 1$) and pericentre passage at time $t_0 = 0$ can be parametrised in terms of the eccentric anomaly u as

$$r(u) = a(1 - e\cos u), \quad \sqrt{\mu/a} \cdot t(u) = a(u - e\sin u).$$

The relation between a and h is given by $h = -\mu/2a$, see equation (3.8). In the limit $e \to 1$ we obtain the curve

$$r(u) = a(1 - \cos u), \quad \sqrt{\mu/a} \cdot t(u) = a(u - \sin u).$$

This is a cycloid as described in Section 4.2, that is, the curve in the

$(\sqrt{\mu/a} \cdot t, r)$-plane described by a point on the boundary of a disc of radius a rolling along the t-axis.

We want to show that this limit curve describes the motion for $c = 0$. Thus, consider a solution of the Kepler problem with $h < 0$ and $c = 0$. As shown in the proof of Theorem 3.1, this motion is along a line, and we have $e = 1$ in this case. Notice that this is consistent with equation (3.7). So we may think of this solution as a curve $t \mapsto r(t) \in \mathbb{R}^+$ satisfying the one-dimensional Kepler equation $\ddot{r} = -\mu/r^2$.

Write $2a$ for the maximal distance of the body from the origin, where its velocity is zero. By conservation of energy we have again $h = -\mu/2a$, see (3.6). The trajectory can be parametrised as

$$r(u) = a(1 - \cos u), \quad u \in (0, 2\pi).$$

We may assume that u grows with t, i.e. $\dot{u} > 0$. Write $r' := \mathrm{d}r/\mathrm{d}u$. Show the following.

(a) $(r')^2 = 2ar - r^2$.
(b) $r\dot{r} = \sqrt{\mu/a} \cdot r'$.
(c) $\dot{u} = \sqrt{\mu/a} \cdot (1/r)$.
(d) $\sqrt{\mu/a} \cdot t = a(u - \sin u)$, where $t = 0$ is the time of 'emission' from the central body.
(e) The time required for a body to fall to the force centre from rest at $r = r_0$ equals $(r_0/2)^{3/2}\pi/\sqrt{\mu}$.

You will notice that one does not have to appeal to the differential equation for r, but merely to the equation coming from conservation of energy.

Statement (c) implies that in the case $c = 0$, too, the parameter u can be defined as an (improper) integral

$$u(t) = \sqrt{\frac{\mu}{a}} \int_0^t \frac{\mathrm{d}\tau}{r(\tau)}$$

as in the preceding exercise.

So in the limit $e \to 1$ (for given energy $h < 0$) we do indeed obtain the solution for $c = 0$. It is now sensible to allow u to vary over all of $\mathbb{R}$ even in this case $c = 0$. This corresponds to the particle being reflected at the central body in a collision.[5] We then obtain a family of solutions of the Kepler problem depending continuously on $\mathbf{c}$ (resp. $\mathbf{e}$), with $c = 0$ (resp. $e = 1$) included.

[5] This extension of collision solutions to solutions defined for all times is known as **regularisation**. See also Chapter 8.

4.8 Carry out a similar analysis as in the preceding exercise for the case of positive energy.

4.9 Let $t \mapsto r(t) \in \mathbb{R}^+$ be the unique solution of the one-dimensional Kepler problem $\ddot{r} = -\mu/r^2$ with $r(0) = r_0$, positive initial velocity $\dot{r}(0)$ and zero energy, i.e.

$$\frac{1}{2}\dot{r}^2 - \frac{\mu}{r} = 0.$$

Show by integration of this energy equation that the solution is given explicitly as

$$r(t) = \left(3\sqrt{\frac{\mu}{2}}\,t + r_0^{3/2}\right)^{2/3},$$

defined on the appropriate interval (t_0, ∞) where the expression in parentheses is positive.

Alternatively, derive this solution by taking the limit $c \to 0$ in Proposition 4.7.

4.10 The Kepler problem is of order 6, being described by a second-order differential equation $\ddot{\mathbf{r}} = -\mu\mathbf{r}/r^3$ in three spatial variables. In other words, a solution is uniquely determined by six parameters, for instance the initial values $\mathbf{r}(0)$ and $\dot{\mathbf{r}}(0)$. In the astronomy of the solar system, traditionally the six parameters or **orbital elements** described presently are used to determine an orbit.

Fix a coordinate system with the Sun S at the origin. The particular choice of the corresponding orthonormal basis $\mathbf{e}_1, \mathbf{e}_2, \mathbf{e}_3$ for $\mathbb{R}^3$ is irrelevant; in the astronomical practice one chooses $\mathbf{e}_1$ and $\mathbf{e}_2$ to span the orbital plane of the Earth, the so-called **ecliptic**, such that the orbit of the Earth is positively oriented in that plane. The vector $\mathbf{e}_3$ is then chosen to form a right-handed basis with $\mathbf{e}_1$ and $\mathbf{e}_2$. The vector $\mathbf{e}_1$ is chosen to point in the direction of the **vernal equinox**, i.e. the point on the celestial sphere where the Sun is located at the beginning of spring.

Show that the orbit of any other planet, provided it does not lie entirely in the ecliptic, can be determined uniquely (as a parametrised curve) by the following parameters.

(a) The **inclination**, i.e. the angle i between $\mathbf{e}_3$ and the normal vector $\mathbf{n}$ to the orbital plane, where $\mathbf{n}$ is chosen to point in the direction of the angular momentum of the planet.

(b) The **longitude of the ascending node**, i.e. the angle Ω between $\mathbf{e}_1$ and the direction from S to the intersection A of the planetary orbit with the ecliptic, where the planet passes upward through the ecliptic.

(c) The **eccentricity** e of the elliptic orbit.
(d) The **semi-major axis** a of the ellipse.
(e) The **argument of the pericentre** (also called **perihelion** for the Sun being the central body), that is, the angle $\omega = \measuredangle ASP$, where P denotes the perihelion, i.e. the point on the orbit closest to the Sun.
(f) The time t_0 of a **pericentre passage**.

The first two parameters determine the orbital plane, the next three specify the shape and position of the orbit in that plane, and the final one locates the planet on its orbit. Draw a picture to illustrate these orbital elements.

How do the angular momentum $\mathbf{c}$ and the energy h, which are other constants of motion, relate to the orbital elements? More precisely: how can one determine $\mathbf{c}$ and h from the orbital elements? Conversely, given $\mathbf{c}$ and h, which other two quantities are necessary to determine the orbit?

5

The two-body problem

But it takes two to tango.

Hoffman & Manning

We now study the motion of two bodies under mutual gravitational influence. We shall see that the corresponding system of differential equations reduces in two different ways to the Kepler problem.

Let m_1, m_2 be the masses of the two bodies, and $\mathbf{r}_1, \mathbf{r}_2$ their position vectors relative to an arbitrary coordinate system. Then the equations for the two-body problem are

$$m_1 \ddot{\mathbf{r}}_1 = G m_1 m_2 \frac{\mathbf{r}_2 - \mathbf{r}_1}{|\mathbf{r}_2 - \mathbf{r}_1|^3}, \tag{5.1}$$

$$m_2 \ddot{\mathbf{r}}_2 = G m_1 m_2 \frac{\mathbf{r}_1 - \mathbf{r}_2}{|\mathbf{r}_1 - \mathbf{r}_2|^3}. \tag{5.2}$$

A solution of the two-body problem is a C^2-map

$$(\mathbf{r}_1, \mathbf{r}_2)\colon\ I \to (\mathbb{R}^3 \times \mathbb{R}^3) \setminus \Delta$$

satisfying these equations, where $I \subset \mathbb{R}$ is some interval, and

$$\Delta := \{(\mathbf{x}, \mathbf{x})\colon\ \mathbf{x} \in \mathbb{R}^3\}$$

is the diagonal in $\mathbb{R}^3 \times \mathbb{R}^3$.

As in Exercise 1.1, it is easy to see that solutions $(\mathbf{r}_1, \mathbf{r}_2)$ of the two-body problem have the following invariance properties.

(i) Invariance under isometries, i.e. distance-preserving maps: if $A\colon\ \mathbb{R}^3 \to \mathbb{R}^3$ is an isometry, then $(A\mathbf{r}_1, A\mathbf{r}_2)$ is also a solution.
(ii) Galilean relativity principle: $(\mathbf{r}_1 + \mathbf{a}t + \mathbf{b}, \mathbf{r}_2 + \mathbf{a}t + \mathbf{b})$ is a solution for any $\mathbf{a}, \mathbf{b} \in \mathbb{R}^3$.

5.1 Reduction to relative coordinates

Set $\mathbf{r} := \mathbf{r}_2 - \mathbf{r}_1$. Divide (5.1) by m_1 and (5.2) by m_2, then subtract the first equation from the second equation. This yields

$$\ddot{\mathbf{r}} = -G(m_1 + m_2)\,\frac{\mathbf{r}}{r^3},$$

i.e. the Kepler problem with $\mu = G(m_1 + m_2)$. Once $\mathbf{r}$ is known, one finds $\mathbf{r}_1$ from (5.1) by integration (with integration constants $\mathbf{a}, \mathbf{b}$ as in the Galilean relativity principle), and then $\mathbf{r}_2$ as $\mathbf{r}_2 = \mathbf{r} + \mathbf{r}_1$.

5.2 Reduction to barycentric coordinates

The **centre of mass** of the two-body system is given by

$$\boxed{\mathbf{r}_c := \frac{m_1\mathbf{r}_1 + m_2\mathbf{r}_2}{m_1 + m_2}.}$$

Notice that this point lies on the line segment joining $\mathbf{r}_1$ with $\mathbf{r}_2$.

We compute, again with $\mathbf{r} := \mathbf{r}_2 - \mathbf{r}_1$,

$$\begin{aligned}\ddot{\mathbf{r}}_c &= \frac{m_1}{m_1 + m_2}\,\ddot{\mathbf{r}}_1 + \frac{m_2}{m_1 + m_2}\,\ddot{\mathbf{r}}_2 \\ &= \frac{Gm_1m_2}{m_1 + m_2}\left(\frac{\mathbf{r}}{r^3} - \frac{\mathbf{r}}{r^3}\right) \\ &= \mathbf{0}.\end{aligned}$$

By integrating we obtain $\mathbf{r}_c(t) = \mathbf{a}t + \mathbf{b}$ for some $\mathbf{a}, \mathbf{b} \in \mathbb{R}^3$.

The two-body problem is described by a system of second-order differential equations in six spatial variables $\mathbf{r}_1, \mathbf{r}_2$, so it is of order 12. A solution will be determined, for instance, by the values $\mathbf{r}_1(0), \dot{\mathbf{r}}_1(0), \mathbf{r}_2(0), \dot{\mathbf{r}}_2(0)$. The vectors $\mathbf{a}, \mathbf{b}$ give six constants of motion, so the remaining system should be of order 6. And indeed this is exactly what happens when we reduce again to a Kepler problem relative to $\mathbf{r}_c$.

Set $\tilde{\mathbf{r}}_1 := \mathbf{r}_1 - \mathbf{r}_c$ and $\tilde{\mathbf{r}}_2 := \mathbf{r}_2 - \mathbf{r}_c$. Because of $\ddot{\mathbf{r}}_c = \mathbf{0}$, these position vectors relative to the centre of mass still satisfy the equations of the two-body problem (5.1) and (5.2).

The new centre of mass, as one should expect, is now fixed at $\mathbf{0}$:

$$\tilde{\mathbf{r}}_c = \frac{m_1\tilde{\mathbf{r}}_1 + m_2\tilde{\mathbf{r}}_2}{m_1 + m_2} = \frac{m_1\mathbf{r}_1 + m_2\mathbf{r}_2}{m_1 + m_2} - \mathbf{r}_c = \mathbf{0}.$$

In other words, it suffices to solve the two-body problem under the additional assumption that the centre of mass be fixed at the origin:

$$m_1\mathbf{r}_1 + m_2\mathbf{r}_2 = \mathbf{0}. \tag{5.3}$$

If $\mathbf{r}_1, \mathbf{r}_2$ is a solution of the two-body problem subject to this constraint on the centre of mass, the general solution will be of the form

$$(\mathbf{r}_1 + \mathbf{a}t + \mathbf{b}, \mathbf{r}_2 + \mathbf{a}t + \mathbf{b}).$$

With (5.3) we have $\mathbf{r}_2 = -m_1\mathbf{r}_1/m_2$. This reduces (5.1) to

$$\begin{aligned}\ddot{\mathbf{r}}_1 &= Gm_2\,\frac{-(m_1/m_2+1)\,\mathbf{r}_1}{(m_1/m_2+1)^3 r_1^3}\\ &= -\frac{Gm_2^3}{(m_1+m_2)^2}\cdot\frac{\mathbf{r}_1}{r_1^3}.\end{aligned}$$

Similarly, we have

$$\ddot{\mathbf{r}}_2 = -\frac{Gm_1^3}{(m_1+m_2)^2}\cdot\frac{\mathbf{r}_2}{r_2^3}.$$

Because of (5.3), the solution of one of these differential equations determines that of the other.

Thus, in the two-body problem the centre of mass moves uniformly along a straight line, and the two bodies move relative to the centre of mass as in the Kepler problem, with $\mu = Gm_2^3/(m_1+m_2)^2$ for the first body, and $\mu = Gm_1^3/(m_1+m_2)^2$ for the second. Because of $\mathbf{r}_2 = -m_1\mathbf{r}_1/m_2$, the two orbits lie in the same plane through the centre of mass.

Exercises

5.1 Two planets of respective masses m_1, m_2 are moving on elliptic orbits around a heavy star of mass m_0. Idealising a little, we regard the motions of the two planets as separate two-body problems with the central star. Let a_i and p_i, $i = 1, 2$, be the semi-major axis and period, respectively, of the planetary orbit relative to the star. Determine the ratio

$$\frac{p_1^2/a_1^3}{p_2^2/a_2^3}.$$

This gives a version of Kepler's third law for several planets.

5.2 (a) Describe an explicit solution of the two-body problem with two bodies P_1, P_2 of respective masses m_1 and m_2 moving on circular orbits around the common centre of mass. To this end, describe the orbits as parametrised curves $t \mapsto \mathbf{r}_i(t)$, $i = 1, 2$, and verify the differential equations of the two-body problem. What is the relation between the radii of the two circular orbits in terms of m_1 and m_2?

(b) Explain how the motion of P_1 relative to P_2, the motion of P_1 relative to the centre of mass, and that of P_2 relative to the centre of mass compare as regards

(1) orbit type (elliptic, parabolic, hyperbolic),
(2) eccentricity of the orbit,
(3) semi-major axis in the elliptic case, distance to the directrix in the parabolic case, and real semi-axis in the hyperbolic case.

(c) Assume that the centre of mass is fixed at the origin. Set $\mathbf{v}_i = \dot{\mathbf{r}}_i$, $i = 1, 2$, and define the energy of the two bodies by

$$E_1 = \frac{1}{2} m_1 v_1^2 - \frac{G m_1 m_2^3}{(m_1 + m_2)^2} \cdot \frac{1}{r_1} =: T_1 + V_1$$

and

$$E_2 = \frac{1}{2} m_2 v_2^2 - \frac{G m_2 m_1^3}{(m_1 + m_2)^2} \cdot \frac{1}{r_2} =: T_2 + V_2,$$

respectively. Notice that E_1 and E_2 are constant by Proposition 1.8. Determine the ratio T_1/T_2 between the kinetic energies, the ratio V_1/V_2 between the potential energies, and the ratio E_1/E_2 between the total energies.

6

The n-body problem

> Had the title not been already preempted, one might suggest that the study of the N-body problem is 'the world's oldest profession.' If it isn't the oldest, then, most surely, it is 'the second oldest.'
>
> Donald Saari (1990)

We now consider $n \geq 2$ bodies of respective masses $m_1, \ldots, m_n$ that move in $\mathbb{R}^3$ under the influence of mutual gravitational attraction. Write $\mathbf{r}_1, \ldots, \mathbf{r}_n$ for the position vectors of these bodies. The system of equations for the n-body problem then becomes

$$m_i \ddot{\mathbf{r}}_i = \sum_{\substack{j=1 \\ j \neq i}}^{n} G m_i m_j \frac{\mathbf{r}_j - \mathbf{r}_i}{|\mathbf{r}_j - \mathbf{r}_i|^3}, \quad i = 1, \ldots, n. \tag{6.1}$$

A solution of this n-body problem is a C^2-map

$$t \longmapsto (\mathbf{r}_1(t), \ldots, \mathbf{r}_n(t)) \in \mathbb{R}^{3n} \setminus \Delta,$$

defined on some interval in $\mathbb{R}$, where Δ is the 'thick' diagonal:

$$\Delta := \bigcup_{1 \leq i < j \leq n} \Delta_{ij},$$

with

$$\Delta_{ij} := \{(\mathbf{x}_1, \ldots, \mathbf{x}_n) \in \mathbb{R}^{3n} \colon \; \mathbf{x}_i = \mathbf{x}_j\}.$$

The Δ_{ij} are linear subspaces of $\mathbb{R}^{3n}$ and in particular closed as topological subsets. So Δ is likewise closed, and $\mathbb{R}^{3n} \setminus \Delta$ is open.

In this chapter we deal with some fundamental aspects of the n-body problem. In Section 6.1 we show that the n-body problem can be written as a conservative system, so in particular we have conservation of energy. The homogeneity of the potential is used to show that equilibrium solutions do not exist. In Section 6.2 we discuss under what conditions solutions of the n-body problem exist in finite time only. In Section 6.3 we introduce the moment of inertia and prove the so-called Lagrange–Jacobi identity. In Section 6.4 we meet two further constants of motion (besides the energy): the linear and the angular momentum. In Section 6.5 we prove Sundman's inequality relating energy, angular momentum and moment of inertia, and use this to prove Sundman's theorem on n-body systems that collapse to a single point.

Section 6.6 is concerned with so-called central configurations; these are n-body configurations that collapse homothetically to a point from a rest position. We relate this to general homographic solutions, where the configuration stays self-similar for all times. For three bodies, such homographic solutions form the content of a famous theorem due to Lagrange, which we shall discuss in the next chapter.

6.1 The Newton potential

Definition 6.1 The **Newton potential** is the smooth function

$$V\colon \mathbb{R}^{3n} \setminus \Delta \longrightarrow \mathbb{R}^-$$

defined by

$$\boxed{V(\mathbf{r}_1, \ldots, \mathbf{r}_n) := -\sum_{1\leq i<j\leq n} \frac{Gm_i m_j}{|\mathbf{r}_i - \mathbf{r}_j|}.}$$

The following lemma shows that the force field on $\mathbb{R}^{3n} \setminus \Delta$ describing the n-body problem is conservative, cf. Section 1.2. I first explain the notation. If $V = V(\mathbf{x}, \mathbf{y}, \ldots)$ is a C^1-function of vector variables $\mathbf{x} = (x_1, x_2, x_3), \ldots,$ we set

$$\frac{\partial V}{\partial \mathbf{x}} = \left(\frac{\partial V}{\partial x_1}, \frac{\partial V}{\partial x_2}, \frac{\partial V}{\partial x_3}\right),$$

that is, $\partial V/\partial \mathbf{x}$ are the components of the gradient of V corresponding to $\mathbf{x}$.

Lemma 6.2 *The Newton potential V satisfies*

$$-\frac{\partial V}{\partial \mathbf{r}_i} = \sum_{\substack{j=1\\ j\neq i}}^{n} Gm_i m_j \frac{\mathbf{r}_j - \mathbf{r}_i}{|\mathbf{r}_j - \mathbf{r}_i|^3}.$$

Proof For vector variables $(\mathbf{x}, \mathbf{y}) \in \mathbb{R}^3 \times \mathbb{R}^3$ we have

$$\frac{\partial}{\partial x_k}|\mathbf{x} - \mathbf{y}| = \frac{\partial}{\partial x_k}\sqrt{\langle \mathbf{x} - \mathbf{y}, \mathbf{x} - \mathbf{y}\rangle} = \frac{x_k - y_k}{\sqrt{\langle \mathbf{x} - \mathbf{y}, \mathbf{x} - \mathbf{y}\rangle}},$$

hence

$$\frac{\partial}{\partial \mathbf{x}}|\mathbf{x} - \mathbf{y}| = \frac{\mathbf{x} - \mathbf{y}}{|\mathbf{x} - \mathbf{y}|}.$$

A similar computation shows that

$$\frac{\partial}{\partial \mathbf{x}}\frac{1}{|\mathbf{x} - \mathbf{y}|} = -\frac{1}{|\mathbf{x} - \mathbf{y}|^2} \cdot \frac{\mathbf{x} - \mathbf{y}}{|\mathbf{x} - \mathbf{y}|}.$$

When we apply this to V, the claimed formula follows. □

So the equations of the n-body problem can indeed be written as a conservative system:

$$m_i\ddot{\mathbf{r}}_i = -\frac{\partial V}{\partial \mathbf{r}_i}, \quad i = 1, \ldots, n.$$

The **kinetic energy** of a solution $(\mathbf{r}_1, \ldots, \mathbf{r}_n)$ of the n-body problem is

$$T(t) := \frac{1}{2}\sum_{i=1}^{n} m_i v_i^2(t),$$

where $\mathbf{v}_i(t) := \dot{\mathbf{r}}_i(t)$. As in Proposition 1.8, we have conservation of the total energy.

Proposition 6.3 *For a solution* $\mathbf{r}$ *of the n-body problem, the* **total energy**

$$E(t) := T(t) + V(\mathbf{r}(t))$$

is constant.

Proof We compute

$$\begin{aligned}\frac{\mathrm{d}E}{\mathrm{d}t} &= \sum_{i=1}^{n} m_i\langle \ddot{\mathbf{r}}_i, \dot{\mathbf{r}}_i\rangle + \langle \operatorname{grad} V(\mathbf{r}), \dot{\mathbf{r}}\rangle \\ &= \sum_{i=1}^{n} m_i\langle \ddot{\mathbf{r}}_i, \dot{\mathbf{r}}_i\rangle + \sum_{i=1}^{n}\left\langle \frac{\partial V}{\partial \mathbf{r}_i}, \dot{\mathbf{r}}_i\right\rangle \\ &= 0,\end{aligned}$$

so E is constant. □

Notation 6.4 We write h for the constant total energy of a given solution $\mathbf{r}$, so that we have

$$\boxed{h = T + V(\mathbf{r}).}$$

Is it possible to distribute n bodies in space in such a way that all the gravitational forces acting on each body cancel out, i.e. is there a constant or so-called **equilibrium solution** of the n-body problem? The following proposition answers this question in the negative.

Proposition 6.5 *The n-body problem does not have any equilibrium solutions.*

Heuristically, this result is quite obvious. Given any configuration of n bodies, one can find a 2-plane through one of these bodies such that all other bodies lie in one of the open half-spaces determined by this plane. The resulting force on the first body would have a non-zero component pointing into that half-space.

The key to proving this result is a certain homogeneity property of the Newton potential. Here are the relevant definitions.

Definition 6.6 A subset $\Omega \subset \mathbb{R}^d$ is called a **cone** if for any $\mathbf{x} \in \Omega$ and $\lambda \in \mathbb{R}^+$ the point $\lambda\mathbf{x}$ is likewise in Ω.

Example 6.7 The subset $\mathbb{R}^{3n} \setminus \Delta \subset \mathbb{R}^{3n}$ is a cone.

Definition 6.8 A function $f\colon \Omega \to \mathbb{R}$ on a cone $\Omega \subset \mathbb{R}^d$ is called **positive homogeneous of degree** $p \in \mathbb{R}$ if

$$f(\lambda\mathbf{x}) = \lambda^p f(\mathbf{x}) \quad \text{for all } \mathbf{x} \in \Omega \text{ and } \lambda \in \mathbb{R}^+.$$

Example 6.9 The Newton potential V is positive homogeneous of degree -1.

The following result is attributed to Euler.

Proposition 6.10 *Let $\Omega \subset \mathbb{R}^d$ be an open cone and $f\colon \Omega \to \mathbb{R}$ a C^1-function that is positive homogeneous of degree p. Then*

$$\langle \operatorname{grad} f(\mathbf{x}), \mathbf{x} \rangle = p\, f(\mathbf{x}) \quad \textit{for all } \mathbf{x} \in \Omega.$$

Proof For given $\mathbf{x} \in \Omega$, consider the C^1-function

$$\begin{array}{ccc} \mathbb{R}^+ & \longrightarrow & \mathbb{R} \\ \lambda & \longmapsto & f(\lambda\mathbf{x}) = \lambda^p f(\mathbf{x}). \end{array}$$

When we take the derivative with respect to λ of the two expressions of this function, we obtain

$$\langle \operatorname{grad} f(\lambda\mathbf{x}), \mathbf{x} \rangle = p\lambda^{p-1} f(\mathbf{x}).$$

Setting $\lambda = 1$ we obtain the claimed formula. □

Corollary 6.11 *The Newton potential V satisfies*

$$\sum_{i=1}^{n}\left\langle\frac{\partial V}{\partial \mathbf{r}_i}, \mathbf{r}_i\right\rangle = -V(\mathbf{r}) > 0$$

on $\mathbb{R}^{3n} \setminus \Delta$. □

Proof of Proposition 6.5 Suppose $\mathbf{r}^\circ = (\mathbf{r}_1^\circ, \ldots, \mathbf{r}_n^\circ)$ were an equilibrium solution. Then

$$-\frac{\partial V}{\partial \mathbf{r}_i}(\mathbf{r}^\circ) = m_i\ddot{\mathbf{r}}_i^\circ = \mathbf{0}, \quad i = 1, \ldots, n.$$

In particular, we would have

$$\sum_{i=1}^{n}\left\langle\frac{\partial V}{\partial \mathbf{r}_i}(\mathbf{r}^\circ), \mathbf{r}_i^\circ\right\rangle = 0,$$

contradicting Corollary 6.11. □

6.2 Maximal solutions

In Corollary 3.3 we saw that solutions of the Kepler problem with non-vanishing angular momentum are defined for all times. In the present section we want to characterise solutions of the n-body problem that are defined in finite (positive or negative) time only.

Write the equations (6.1) for the n-body problem as a system of first order:

$$\left.\begin{array}{rcl} \dot{\mathbf{r}}_i & = & \mathbf{v}_i \\ \dot{\mathbf{v}}_i & = & -\dfrac{1}{m_i}\cdot\dfrac{\partial V}{\partial \mathbf{r}_i}(\mathbf{r}) =: \dfrac{1}{m_i}\mathbf{F}_i(\mathbf{r}) \end{array}\right\} \quad i = 1, \ldots, n,$$

where $\mathbf{r} = (\mathbf{r}_1, \ldots, \mathbf{r}_n)$. The right-hand side of this system defines a smooth vector field on $(\mathbb{R}^{3n} \setminus \Delta) \times \mathbb{R}^{3n}$. By the Picard–Lindelöf theorem, for given $\mathbf{r}^0 \in \mathbb{R}^{3n} \setminus \Delta$ and $\mathbf{v}^0 \in \mathbb{R}^{3n}$, the initial value problem

$$\left.\begin{array}{rcl} \mathbf{r}(0) & = & \mathbf{r}^0 \\ \mathbf{v}(0) & = & \mathbf{v}^0 \end{array}\right\}$$

for that system has a unique solution, defined on a maximal time interval (α, ω) with $-\infty \leq \alpha < 0 < \omega \leq \infty$. We want to describe a condition under which ω is finite; the situation $\alpha > -\infty$ is analogous.

Proposition 6.12 *Let $t \mapsto \mathbf{r}(t)$ be a solution of the n-body problem with $\omega < \infty$. Set*

$$\rho(t) = \min\{|\mathbf{r}_i(t) - \mathbf{r}_j(t)|\colon\ 1 \leq i < j \leq n\}, \quad t \in (\alpha, \omega).$$

Then $\rho(t) \to 0$ *for* $t \nearrow \omega$.

Remark 6.13 This statement does not imply that for $t \nearrow \omega$ we must have a collision – only for $n = 2$ and $n = 3$ is this indeed the case (see the notes and references section). What might also happen is that for $t \nearrow \omega$ two bodies go off to infinity with their relative distance tending to zero. However, the proposition does exclude the phenomenon of a single body flying off to infinity in finite time (and the others staying at distances bounded from below).

In preparation for the proof of Proposition 6.12, we need two lemmata.

Lemma 6.14 *Let* $\Omega \subset \mathbb{R}^d$ *be an open subset and* $\mathbf{X}\colon \Omega \to \mathbb{R}^d$ *a locally Lipschitz continuous vector field on* Ω. *Choose* $\mathbf{x}_0 \in \Omega$ *and* $a \in \mathbb{R}^+$ *such that the compact set*[1]

$$\overline{B} := \{\mathbf{x} \in \mathbb{R}^d\colon\ \|\mathbf{x} - \mathbf{x}_0\| \le a\}$$

is contained in Ω. *Set* $K := \max_{\mathbf{x}\in\overline{B}} \|\mathbf{X}(\mathbf{x})\|$. *Then, for* $K > 0$, *the solution of the initial value problem*

$$\dot{\mathbf{x}} = \mathbf{X}(\mathbf{x}), \quad \mathbf{x}(0) = \mathbf{x}_0$$

is defined at least on the interval[2] $[-a/K, a/K]$.

Proof Let (α, ω) be the maximal solution interval. We want to show that $\omega > a/K$. The proof of $\alpha < -a/K$ is analogous.

If $\omega = \infty$, there is nothing to prove. If $\omega < \infty$, then $\mathbf{x}(t)$ cannot lie in $\overline{B}$ for all $t \in [0, \omega)$, see the argument in the proof of Corollary 3.3, and so there has to be a first time when the curve $t \mapsto \mathbf{x}(t)$ reaches the boundary of $\overline{B}$. Set

$$t_1 := \min\{t \in (0, \omega)\colon\ \|\mathbf{x}(t) - \mathbf{x}_0\| = a\},$$

see Figure 6.1.

Heuristically it is clear that $t_1 \ge a/K$: if you always travel at a speed less than K, it takes time at least a/K to travel a distance a. Here is the formal argument. By integrating the differential equation we obtain

$$\mathbf{x}(t) = \mathbf{x}_0 + \int_0^t \mathbf{X}(\mathbf{x}(s))\,\mathrm{d}s,$$

and hence

$$a = \|\mathbf{x}(t_1) - \mathbf{x}_0\| = \left\|\int_0^{t_1} \mathbf{X}(\mathbf{x}(s))\,\mathrm{d}s\right\| \le \int_0^{t_1} \left\|\mathbf{X}(\mathbf{x}(s))\right\| \mathrm{d}s \le K t_1.$$

We conclude that $\omega > t_1 \ge a/K$. □

[1] Here $\|\,.\,\|$ can be any norm on $\mathbb{R}^d$, so $\overline{B}$ need not be a euclidean ball.
[2] For $K = 0$ the solution is obviously constant and defined for all times.

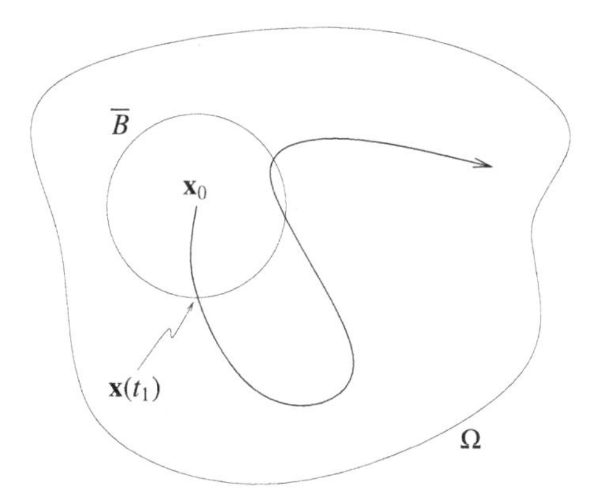

Figure 6.1 A solution leaving $\overline{B}$.

Lemma 6.15 *Given $h \in \mathbb{R}$ and $\delta \in \mathbb{R}^+$, there is a positive real number $\tau = \tau(h, \delta)$ such that any solution of the n-body problem (with given masses $m_1, \ldots, m_n$) of energy h and with initial minimal pairwise distance $\rho(0) \geq \delta$ is defined at least on the time interval $[-\tau, \tau]$.*

Assuming this second lemma, we can now give a short proof of the proposition.

Proof of Proposition 6.12 We argue by contradiction. Suppose $\omega < \infty$, but $\rho(t) \nrightarrow 0$ for $t \nearrow \omega$. Then there is a sequence $t_\nu \nearrow \omega$ and a $\delta > 0$ such that $\rho(t_\nu) \geq \delta$ for all $\nu \in \mathbb{N}$.

For ν sufficiently large we have $t_\nu + \tau(h, \delta) > \omega$, which implies that the solution can be extended beyond the time $t = \omega$, in contradiction to the maximality of ω. □

It remains to prove the second lemma.

Proof of Lemma 6.15 We want to apply Lemma 6.14 to the vector field $\mathbf{X}$ on $\Omega := (\mathbb{R}^{3n} \setminus \Delta) \times \mathbb{R}^{3n}$ defined by

$$\mathbf{X}(\mathbf{r}, \mathbf{v}) := (\mathbf{v}, \mathbf{F}_1/m_1, \ldots, \mathbf{F}_n/m_n) \in \mathbb{R}^{3n} \times \mathbb{R}^{3n}.$$

Consider a solution $t \mapsto (\mathbf{r}(t), \mathbf{v}(t))$ with initial value $\mathbf{x}_0 := (\mathbf{r}(0), \mathbf{v}(0))$, energy h and $\rho(0) \geq \delta$.

Let $|\,.\,|_\infty$ denote the norm on $\mathbb{R}^{3n} \times \mathbb{R}^{3n}$ defined by

$$|(\mathbf{r}, \mathbf{v})|_\infty := \max\{r_1, \ldots, r_n, v_1, \ldots, v_n\},$$

and set

$$\overline{B} := \{(\mathbf{r}, \mathbf{v}) \in \mathbb{R}^{3n} \times \mathbb{R}^{3n} \colon \ |(\mathbf{r}, \mathbf{v}) - \mathbf{x}_0|_\infty \leq \delta/4\}.$$

The triangle inequality gives

$$\delta \le |\mathbf{r}_i(0) - \mathbf{r}_j(0)| \le |\mathbf{r}_i(0) - \mathbf{r}_i| + |\mathbf{r}_i - \mathbf{r}_j| + |\mathbf{r}_j - \mathbf{r}_j(0)|.$$

Hence, for $(\mathbf{r}, \mathbf{v}) \in \overline{B}$ we have

$$|\mathbf{r}_i - \mathbf{r}_j| \ge \delta - |\mathbf{r}_i(0) - \mathbf{r}_i| - |\mathbf{r}_j - \mathbf{r}_j(0)| \ge \delta - \frac{\delta}{4} - \frac{\delta}{4} = \frac{\delta}{2}.$$

In particular, we have $\overline{B} \subset \Omega$.

Our aim now is to estimate the euclidean norm of the components of $\mathbf{v}_i$ and $\mathbf{F}_i/m_i$ of the vector field $\mathbf{X}$ from above by a quantity K depending only on h and δ. Since the 'radius' a of $\overline{B}$ was defined to be $\delta/4$, the term $\tau := a/K$ will then likewise depend on h and δ only, so that the present lemma will be a consequence of Lemma 6.14.

The term $\mathbf{F}_i/m_i$ can be estimated on $\overline{B}$ by

$$|\mathbf{F}_i/m_i| = \left| \sum_{\substack{j=1 \\ j\neq i}}^{n} Gm_j \frac{\mathbf{r}_j - \mathbf{r}_i}{|\mathbf{r}_j - \mathbf{r}_i|^3} \right| \le \sum_{\substack{j=1 \\ j\neq i}}^{n} \frac{Gm_j}{|\mathbf{r}_j - \mathbf{r}_i|^2} \le \frac{4G}{\delta^2} \sum_{\substack{j=1 \\ j\neq i}}^{n} m_j.$$

For $\mathbf{v}_i$ we have this estimate on $\overline{B}$:

$$|\mathbf{v}_i| \le |\mathbf{v}_i - \mathbf{v}_i(0)| + |\mathbf{v}_i(0)| \le \frac{\delta}{4} + |\mathbf{v}_i(0)|,$$

so it remains only to estimate $v_i(0)$. From the energy equality

$$h = \frac{1}{2} \sum_{i=1}^{n} m_i v_i^2(0) + V(\mathbf{r}(0))$$

we have

$$v_i(0) \le \sqrt{\frac{2}{m_i}\big(h - V(\mathbf{r}(0))\big)} \le \sqrt{\frac{2}{m_i}\left(h + \sum_{1\le j<k\le n} \frac{Gm_j m_k}{\delta}\right)}.$$

This completes the proof. □

6.3 The Lagrange–Jacobi identity

Let $t \mapsto \mathbf{r}(t) \in \mathbb{R}^{3n} \setminus \Delta$ be a solution of the n-body problem.

Definition 6.16 The **moment of inertia** of $\mathbf{r}$ is

$$I(t) := \frac{1}{2} \sum_{i=1}^{n} m_i r_i^2(t).$$

Remarks 6.17 (1) Strictly speaking, a physicist would call $2I$ the moment of inertia. In the literature on the n-body problem, I is traditionally defined as above. Some authors correctly refer to $2I$ as the moment of inertia; I follow the slight abuse of language common in the literature.

(2) The expression $\sqrt{\sum_{i=1}^n m_i r_i^2/2}$ defines a norm on $\mathbb{R}^{3n}$. So we can think of I as the norm squared of the spatial component of $t \mapsto (\mathbf{r}(t), \mathbf{v}(t))$, just as the kinetic energy T may be thought of as the norm squared of the $\mathbf{v}$-component.

Theorem 6.18 *The moment of inertia and the energy of a solution* $\mathbf{r}$ *of the* n*-body problem are related by the* **Lagrange–Jacobi identity**

$$\boxed{\ddot{I} = 2T + V(\mathbf{r}) = 2h - V(\mathbf{r}) = T + h.}$$

Proof We compute

$$\dot{I} = \sum_{i=1}^n m_i \langle \dot{\mathbf{r}}_i, \mathbf{r}_i \rangle$$

and

$$\begin{aligned}
\ddot{I} &= \sum_{i=1}^n m_i v_i^2 + \sum_{i=1}^n \langle m_i \ddot{\mathbf{r}}_i, \mathbf{r}_i \rangle \\
&= 2T - \sum_{i=1}^n \left\langle \frac{\partial V}{\partial \mathbf{r}_i}, \mathbf{r}_i \right\rangle \\
&= 2T + V(\mathbf{r}),
\end{aligned}$$

see Proposition 6.10 or the proof of Proposition 6.5. The variants of the identity follow from the definition of the total energy as $h = T + V(\mathbf{r})$. □

Example 6.19 Here is a simple application of this identity. I claim that if $\mathbf{r}\colon (\alpha, \infty) \to \mathbb{R}^{3n} \setminus \Delta$ is a solution of the n-body problem with $h > 0$, then $\lim_{t\to\infty} I(t) = \infty$.

Indeed, we have $\ddot{I} = 2h - V(\mathbf{r}) > 2h$. Choose $t_0 > \alpha$. The fundamental theorem of calculus gives, for $t \geq t_0$,

$$\dot{I}(t) > \dot{I}(t_0) + 2h(t - t_0)$$

and

$$I(t) > I(t_0) + \dot{I}(t_0)(t - t_0) + h(t - t_0)^2,$$

which implies the claim.

We may not conclude, however, that one of the $r_i(t)$ has to go to infinity for $t \to \infty$. It might well happen that an $\mathbf{r}_i$ and an $\mathbf{r}_j$ oscillate in such a way that $r_i^2 + r_j^2 \to \infty$.

6.4 Conservation of momentum

As in Section 5.2 we define the **centre of mass** of the n-body system by

$$\boxed{\mathbf{r}_c := \frac{\sum_{i=1}^n m_i \mathbf{r}_i}{\sum_{i=1}^n m_i}.}$$

As in that section one computes $\ddot{\mathbf{r}}_c = 0$, hence $\mathbf{r}_c(t) = \mathbf{a}t + \mathbf{b}$. The vectors $\mathbf{a}, \mathbf{b}$ give us six constants of motion. The conserved quantity $\sum_{i=1}^n m_i \dot{\mathbf{r}}_i = \mathbf{a} \sum_{i=1}^n m_i$ is called the **linear momentum** of the system.

By passing from $\mathbf{r}_i$ to $\mathbf{r}_i - \mathbf{r}_c$ one may always assume that the centre of mass is fixed at the origin. Notice that $\mathbf{r}_c$ lies in the convex hull of the $\mathbf{r}_i$, see Exercise 6.6.

Definition 6.20 The **angular momentum** of the system $\mathbf{r}$ is

$$\boxed{\mathbf{c} := \sum_{i=1}^n m_i \mathbf{r}_i \times \dot{\mathbf{r}}_i.}$$

As in Proposition 1.3 we see that $\mathbf{c}$ is constant.

Proposition 6.21 *The angular momentum is a constant of motion in the n-body problem.*

Proof We compute

$$\begin{aligned}
\frac{\mathrm{d}\mathbf{c}}{\mathrm{d}t} &= \sum_{i=1}^n m_i (\dot{\mathbf{r}}_i \times \dot{\mathbf{r}}_i + \mathbf{r}_i \times \ddot{\mathbf{r}}_i) \\
&= \sum_{i=1}^n \mathbf{r}_i \times \sum_{\substack{j=1 \\ j \neq i}}^n G m_i m_j \frac{\mathbf{r}_j - \mathbf{r}_i}{|\mathbf{r}_j - \mathbf{r}_i|^3} \\
&= \sum_i \sum_{j \neq i} \frac{G m_i m_j}{|\mathbf{r}_j - \mathbf{r}_i|^3} \mathbf{r}_i \times \mathbf{r}_j,
\end{aligned}$$

which equals zero, since $\mathbf{r}_i \times \mathbf{r}_j$ is skew-symmetric in i and j, whereas the coefficient is symmetric in i and j. □

Remark 6.22 In the two-body problem we found 12 constants of motion: the six given by $\mathbf{a}$ and $\mathbf{b}$, plus six further constants corresponding to a solution of the Kepler problem, such as the orbital elements described in Exercise 4.10. In the general n-body problem we have but 10 constants of motion: $\mathbf{a}, \mathbf{b}, \mathbf{c}$ and h.

6.5 Sundman's theorem on total collapse

Let $\mathbf{r} = (\mathbf{r}_1, \ldots, \mathbf{r}_n)\colon (\alpha, \omega) \to \mathbb{R}^{3n} \setminus \Delta$ be a solution of the n-body problem with centre of mass fixed at the origin. We say the system experiences a **total collapse** if there is a point $\mathbf{p}_\omega \in \mathbb{R}^3$ such that

$$\lim_{t \nearrow \omega} \mathbf{r}_i(t) = \mathbf{p}_\omega, \quad i = 1, \ldots, n.$$

Sundman's theorem says that a total collapse can occur only if the system has zero angular momentum. We first need a couple of lemmata.

Lemma 6.23 *If a total collapse happens, then $\mathbf{p}_\omega = \mathbf{0}$ and $\omega < \infty$.*

Proof If $\mathbf{r}_i(t) \to \mathbf{p}_\omega$ for $t \nearrow \omega$ and $i = 1, \ldots, n$, then also

$$\mathbf{0} = \mathbf{r}_c = \frac{\sum_i m_i \mathbf{r}_i(t)}{\sum_i m_i} \longrightarrow \mathbf{p}_\omega \quad \text{for } t \nearrow \omega.$$

So a total collapse is equivalent to $I(t) \to 0$.

In order to prove $\omega < \infty$, we observe that in a total collapse into the origin we have

$$V(\mathbf{r}(t)) = -G \sum_{i<j} \frac{m_i m_j}{|\mathbf{r}_i(t) - \mathbf{r}_j(t)|} \longrightarrow -\infty$$

for $t \nearrow \omega$. This limit, together with the Lagrange–Jacobi identity, gives

$$\ddot{I}(t) = 2h - V(\mathbf{r}(t)) \longrightarrow \infty.$$

This implies the existence of a $t_0 \in (\alpha, \omega)$ such that $\ddot{I}(t) \geq 1$ for all $t \in [t_0, \omega)$. Applying the fundamental theorem of calculus twice as in Example 6.19, we find that

$$I(t) \geq I(t_0) + \dot{I}(t_0)(t - t_0) + \frac{1}{2}(t - t_0)^2 \quad \text{for } t \geq t_0.$$

If ω were infinite, this would yield $I(t) \to \infty$ as $t \nearrow \omega$, contradicting our previous observation that $I(t) \to 0$. □

Lemma 6.24 (Sundman's inequality) *For any solution of the n-body problem we have*

$$c^2 \leq 4I(\ddot{I} - h).$$

Proof From $\mathbf{c} = \sum_i m_i \mathbf{r}_i \times \dot{\mathbf{r}}_i$ we have

$$c \leq \sum_i m_i |\mathbf{r}_i \times \dot{\mathbf{r}}_i| \leq \sum_i m_i r_i v_i = \sum_i (\sqrt{m_i}\, r_i)(\sqrt{m_i}\, v_i).$$

Applying the Cauchy–Schwarz inequality we obtain

$$c \leq \Big(\sum_i m_i r_i^2\Big)^{1/2} \cdot \Big(\sum_i m_i v_i^2\Big)^{1/2} = \sqrt{2I} \cdot \sqrt{2T}.$$

Now conclude with the Lagrange–Jacobi identity $\ddot{I} = T + h$. □

Remark 6.25 A stronger version of this inequality will be proved in Exercise 6.9. The case of equality in either version of the inequality will be discussed in Exercise 6.10.

Theorem 6.26 (Sundman's theorem on total collapse) *If a total collapse occurs in an n-body system with centre of mass fixed at the origin, then the angular momentum is zero.*

Proof If a total collapse occurs at time ω, which is finite by Lemma 6.23, then $I(t) \to 0$ and, as we saw in the proof of that lemma, $\ddot{I}(t) \to \infty$ as $t \nearrow \omega$. In particular, we have $\ddot{I} > 0$ near ω, so $\dot{I}$ is strictly monotone increasing there. It follows that there is an interval $[t_0, \omega)$ on which $\dot{I}$ does not change sign. But $I > 0$ and $I(t) \to 0$, so we must have $\dot{I} < 0$ on $[t_0, \omega)$. We may choose t_0 so large that $I(t) < 1$ on that interval.

Now multiply both sides of Sundman's inequality by $-\dot{I}/I > 0$ to obtain

$$-c^2 \cdot \frac{\dot{I}}{I} \leq 4(h\dot{I} - \dot{I} \cdot \ddot{I}) \text{ on } [t_0, \omega).$$

Integrate this inequality to get

$$-c^2 \log I \leq 4hI - 2\dot{I}^2 + K \leq 4hI + K \text{ on } [t_0, \omega),$$

where K is a constant of integration. Observing that $-\log I > 0$ on $[t_0, \omega)$, we arrive at the estimate

$$c^2 \leq \frac{4hI + K}{-\log I} \text{ on } [t_0, \omega).$$

As $t \nearrow \omega$, the moment of inertia I tends to zero, hence $-\log I \to \infty$. We conclude that $c = 0$. □

6.6 Central configurations

One way to find solutions of the n-body problem is to impose additional geometric constraints. As we saw in Proposition 6.5, there are no solutions where the configuration of the n bodies remains unchanged over time. The next simplest solutions one might look for are those where the configuration changes only by homotheties. In other words, taking the origin $\mathbf{0} \in \mathbb{R}^3$ as the centre of the homothety, we are looking for solutions $\mathbf{r} = (\mathbf{r}_1, \ldots, \mathbf{r}_n)$ of the form $\mathbf{r}_i(t) = \phi(t)\mathbf{a}_i$ with some fixed $\mathbf{a}_1, \ldots, \mathbf{a}_n \in \mathbb{R}^3$ and a positive real-valued function ϕ.

Plugging this ansatz into the equations (6.1) for the n-body problem gives

$$\phi^2 \ddot{\phi}\, m_i \mathbf{a}_i = \sum_{\substack{j=1 \\ j \neq i}}^{n} G m_i m_j \frac{\mathbf{a}_j - \mathbf{a}_i}{|\mathbf{a}_j - \mathbf{a}_i|^3}, \quad i = 1, \ldots, n. \tag{6.2}$$

The right-hand side of these equations does not depend on time, so there must be a real constant μ such that

$$\ddot{\phi} = -\frac{\mu}{\phi^2} \tag{6.3}$$

and

$$-\mu \mathbf{a}_i = \sum_{\substack{j=1 \\ j \neq i}}^{n} G m_j \frac{\mathbf{a}_j - \mathbf{a}_i}{|\mathbf{a}_j - \mathbf{a}_i|^3}, \quad i = 1, \ldots, n. \tag{6.4}$$

The requirement that the homothety be centred at the origin forces the centre of mass to be fixed at the origin. Indeed, summing (6.4) over i we obtain $\sum_i m_i \mathbf{a}_i = \mathbf{0}$, and hence $\mathbf{r}_c = \mathbf{0}$.

Definition 6.27 A configuration $(\mathbf{a}_1, \ldots, \mathbf{a}_n) \in \mathbb{R}^{3n} \setminus \Delta$ of n given bodies of respective masses $m_1, \ldots, m_n$ is called a **central configuration** if it satisfies condition (6.4) for some $\mu \in \mathbb{R}$.

If $(\mathbf{a}_1, \ldots, \mathbf{a}_n)$ is a central configuration, then so is $(\lambda\mathbf{a}_1, \ldots, \lambda\mathbf{a}_n)$ for any $\lambda \in \mathbb{R}^+$, with μ replaced by μ/λ^3 in (6.4). Moreover, equation (6.4) is invariant under the action of the orthogonal group O(3). It follows that any configuration similar to a central configuration, and with centre of mass at the origin, is again a central configuration. Any two central configurations related in this fashion are regarded as equivalent.

Equation (6.3), for $\mu > 0$, is simply the one-dimensional Kepler problem, whose solutions we discussed in Exercises 4.7 to 4.9. As we shall see presently, condition (6.4) will indeed guarantee that $\mu > 0$.[3] Thus, from the definition it is tautological that any central configuration gives rise to a homothetic solution of the n-body problem. Below I shall describe more interesting motions derived from *planar* central configurations.

Here is a useful characterisation of central configurations.

[3] This is clear physically, since for $\mu \leq 0$ the resulting force on each body would be zero or directed away from the origin, which is impossible.

Proposition 6.28 *The n points* $(\mathbf{a}_1, \ldots, \mathbf{a}_n) =: \mathbf{a}$ *form a central configuration if and only if* $\mathbf{a}$ *is a critical point of the function*[4] $V^2 I$.

Proof We have

$$\frac{\partial I}{\partial \mathbf{r}_i} = m_i \mathbf{r}_i,$$

so (6.4) can be rewritten as

$$\mu \frac{\partial I}{\partial \mathbf{r}_i}(\mathbf{a}) = \frac{\partial V}{\partial \mathbf{r}_i}(\mathbf{a}), \quad i = 1, \ldots, n.$$

Since I is homogeneous of degree 2, and V, of degree -1, Proposition 6.10 gives

$$\mu \cdot 2I(\mathbf{a}) = \mu \left\langle \frac{\partial I}{\partial \mathbf{r}}(\mathbf{a}), \mathbf{a} \right\rangle = \left\langle \frac{\partial V}{\partial \mathbf{r}}(\mathbf{a}), \mathbf{a} \right\rangle = -V(\mathbf{a}),$$

that is,

$$\mu = -\frac{V(\mathbf{a})}{2I(\mathbf{a})} > 0.$$

Since $V(\mathbf{a}) \neq 0$, we can write (6.4) equivalently as

$$2V(\mathbf{a}) \frac{\partial V}{\partial \mathbf{r}_i}(\mathbf{a}) I(\mathbf{a}) + V^2(\mathbf{a}) \frac{\partial I}{\partial \mathbf{r}_i}(\mathbf{a}) = 0, \quad i = 1, \ldots, n,$$

which is saying that $\mathbf{a}$ is a critical point of $V^2 I$. □

Clearly, any two-body configuration is central. For three or four bodies, the central configurations among *generic* configurations are easy to characterise.

Example 6.29 Since the centre of mass of a central configuration is fixed at the origin, we can write

$$-\mathbf{a}_i = \frac{1}{M} \sum_{j=1}^{n} m_j (\mathbf{a}_j - \mathbf{a}_i) = \frac{1}{M} \sum_{\substack{j=1 \\ j \neq i}}^{n} m_j (\mathbf{a}_j - \mathbf{a}_i),$$

with $M := m_1 + \cdots + m_n$ the total mass. It follows that (6.4) can be rewritten as

$$\sum_{\substack{j=1 \\ j \neq i}}^{n} m_j \left(\frac{G}{|\mathbf{a}_j - \mathbf{a}_i|^3} - \frac{\mu}{M} \right) \cdot (\mathbf{a}_j - \mathbf{a}_i) = \mathbf{0}, \quad i = 1, \ldots, n.$$

If, for given i, the vectors $\mathbf{a}_j - \mathbf{a}_i$, $j \neq i$, are linearly independent, the coefficients in this equation must vanish. Hence the following statements hold.

[4] With the masses $m_1, \ldots, m_n$ taken as given, both the Newton potential V and the moment of inertia I are functions of $\mathbf{r}$.

(1) Three bodies not on a line form a central configuration if and only if the bodies form an equilateral triangle.
(2) Four bodies not in a plane form a central configuration if and only if the bodies form a regular tetrahedron.

Observe that the constant μ in the corresponding Kepler problem is then given by $\mu = GM/s^3$, where s is the edge length of the triangle or tetrahedron, respectively.

Example 6.30 The regular n-gon is a central configuration for n equal masses (Exercise 6.13).

Given a **planar** central configuration $(\mathbf{a}_1, \dots, \mathbf{a}_n) \in (\mathbb{R}^2)^n$, one can derive more interesting solutions of the n-body problem than homothetic ones as follows. Identify $\mathbb{R}^2$ with $\mathbb{C}$ and start with the ansatz $\mathbf{r}_i(t) = \boldsymbol{\phi}(t)\mathbf{a}_i$, where $\boldsymbol{\phi}$ is now allowed to take values in $\mathbb{C}$. This yields equations like (6.2), but with $\phi^2\ddot{\phi}$ replaced by $|\boldsymbol{\phi}|^3\boldsymbol{\phi}^{-1}\ddot{\boldsymbol{\phi}}$, which leads to the two-dimensional Kepler problem

$$\ddot{\boldsymbol{\phi}} = -\frac{\mu}{|\boldsymbol{\phi}|^2} \cdot \frac{\boldsymbol{\phi}}{|\boldsymbol{\phi}|}.$$

For instance, given any three bodies positioned at the vertices of an equilateral triangle, any solution of the corresponding two-dimensional Kepler problem gives rise to a planar solution of the three-body problem where the three bodies move on similar conic sections and form an equilateral triangle at any given moment. These solutions were discovered by Lagrange, and we shall investigate such solutions in greater detail in the following chapter.

The solutions just described are examples of the following kind.

Definition 6.31 A solution of the n-body problem where the configuration formed by the bodies stays self-similar is called **homographic**.

In other words, what we have seen is that any planar central configuration gives rise to a planar[5] homographic solution. In fact, any planar homographic solution arises in this fashion from a central configuration.

Proposition 6.32 *Let $(\mathbf{r}_1, \dots, \mathbf{r}_n)$ be a planar homographic solution of the n-body problem (with centre of mass fixed at the origin). Then $\mathbf{r}_i = \boldsymbol{\phi}\mathbf{a}_i$, $i = 1, \dots, n$, where $(\mathbf{a}_1, \dots, \mathbf{a}_n)$ is a planar central configuration, and $\boldsymbol{\phi}$ is a solution of the corresponding two-dimensional Kepler problem.*

[5] A solution of the n-body problem is called **planar** if the motion occurs in a fixed plane; it is called **co-planar** if the bodies stay in a common plane that may be rotating in space about the centre of mass.

Proof If we identify the plane of motion with $\mathbb{C} \equiv \mathbb{R}^2 \times \{0\} \subset \mathbb{R}^3$, the planar homographic solution $(\mathbf{r}_1, \dots, \mathbf{r}_n)$ can be written as

$$\mathbf{r}_i(t) = \lambda(t) e^{i\varphi(t)} \mathbf{a}_i, \quad i = 1, \dots, n,$$

where $(\mathbf{a}_1, \dots, \mathbf{a}_n)$ is some planar configuration, and the functions λ and φ take values in $\mathbb{R}^+$ and $\mathbb{R}$, respectively. We compute

$$\dot{\mathbf{r}}_i = (\dot{\lambda} + i\lambda\dot{\varphi})\, e^{i\varphi} \mathbf{a}_i$$

and

$$\ddot{\mathbf{r}}_i = (\ddot{\lambda} - \lambda\dot{\varphi}^2 + i(2\dot{\lambda}\dot{\varphi} + \lambda\ddot{\varphi}))\, e^{i\varphi} \mathbf{a}_i.$$

Since $i\mathbf{a}_i$ is orthogonal to $\mathbf{a}_i$, the angular momentum of the system is

$$\mathbf{c} = \sum_i m_i \mathbf{r}_i \times \dot{\mathbf{r}}_i = (0, 0, \lambda^2 \dot{\varphi}\, \Sigma_i m_i |\mathbf{a}_i|^2).$$

This being constant, we conclude that $\lambda^2\dot{\varphi}$ is constant. Hence $\lambda(2\dot{\lambda}\dot{\varphi} + \lambda\ddot{\varphi})$ vanishes identically and, since $\lambda \neq 0$, so does $2\dot{\lambda}\dot{\varphi} + \lambda\ddot{\varphi}$. Thus, $\ddot{\mathbf{r}}_i$ simplifies to

$$\ddot{\mathbf{r}}_i = (\ddot{\lambda} - \lambda\dot{\varphi}^2)\, e^{i\varphi} \mathbf{a}_i.$$

Plugging these expressions for $\mathbf{r}_i$ and $\ddot{\mathbf{r}}_i$ into the equations for the n-body problem, we obtain

$$\lambda^2(\ddot{\lambda} - \lambda\dot{\varphi}^2)\, \mathbf{a}_i = \sum_{\substack{j=1 \\ j \neq i}}^{n} G m_j \frac{\mathbf{a}_j - \mathbf{a}_i}{|\mathbf{a}_j - \mathbf{a}_i|^3}.$$

These are the equations for a central configuration. By our previous considerations, $\boldsymbol{\phi} = \lambda e^{i\varphi}$ must be a solution of the two-dimensional Kepler problem. □

Remark 6.33 If we set $\gamma := \lambda^2\dot{\varphi}$, which – as we have just seen – is constant, we can write the constant μ for this central configuration as

$$\lambda^2\ddot{\lambda} - \frac{\gamma^2}{\lambda} = -\mu.$$

This happens to be the differential equation for the coordinate r in the Kepler problem, see Exercise 3.2.

A special class of homographic solutions consists of those for which the configuration does not change its shape at all.

Definition 6.34 A solution of the n-body problem where the configuration formed by the bodies stays self-congruent is called a **relative equilibrium**.

Example 6.35 Given any planar central configuration, the circular solutions of the corresponding two-dimensional Kepler problem give rise to relative equilibria.

With the methods we are going to develop in the next chapter, we shall see that relative equilibrium solutions are planar, and the configuration rotates with constant angular velocity (Exercise 7.7). In other words, relative to a rotating frame this looks like an equilibrium solution, which explains the terminology. This also means that all relative equilibrium solutions arise from a planar central configuration as in the preceding example.

Notes and references

In Section 6.2 I followed (Ortega and Ureña, 2010). The remaining parts of this chapter, apart from Section 6.6, can be found in similar form both in that Spanish text and in (Pollard, 1966). Pollard's book contains a number of other applications of the Lagrange–Jacobi identity, such as the so-called virial theorem (deriving from the Latin word *vis* for force or energy), which relates the average over time of the kinetic energy to that of the potential energy. This allows one to make statements about the long-time growth of n-body systems.

The fact, mentioned in Remark 6.13, that solutions of the three-body problem defined in finite time only must experience a collision was established in (Sundman, 1912, page 121). Sundman attributes an earlier proof of this result to Painlevé. The cited memoir contains Sundman's theorem on total collapse and a remark to the effect that this theorem was known previously to Weierstraß, who never published this result, but communicated it in a letter to Mittag-Leffler. The 'constantes des aires' mentioned by Sundman (page 148) in the formulation of the theorem are simply the cartesian components of the total angular momentum. The memoir also includes Sundman's most striking contribution to celestial mechanics, *viz.*, the construction of a power series solution to the three-body problem in the case $c \neq 0$, with collisions suitably regularised.[6] Unfortunately, the slow convergence rate of this infinite series means that it is of little practical value.

For a more advanced introduction to the n-body problem from the viewpoint of Hamiltonian dynamical systems see (Meyer *et al.*, 2009). Among many other things that book contains a chapter on special coordinates, where one can find a general treatment of the so-called Jacobi coordinates (see Exercise 6.11) and alternative proofs of Kepler's first law using special coordinates.

[6] On the issue of regularisation see Exercise 4.7 and Chapter 8.

I quote from that text: "There is an old saying in celestial mechanics that 'no set of coordinates is good enough.' Indeed, classical and modern literatures are replete with endless coordinate changes." Exercise 6.12 is an example showing that a judicious choice of coordinates (here: Jacobi coordinates) may be required for tackling a particular problem. A basic introduction to Hamiltonian systems will be given in Chapter 9.

For more information on central configurations and homographic solutions see (Wintner, 1941), (Saari, 2005), (Moeckel, 2014) and (Moeckel, 2015); these references give ample justification for the significance of central configurations beyond their being pretty. One important question is whether for a given number n of bodies with given masses the number of central configurations is finite. This is a difficult question even for the planar case, i.e. for relative equilibria, and indeed it was listed as one of 18 problems for our century by Smale (1998). The latter question has a positive answer for $n \leq 4$ (Hampton and Moeckel, 2006). The central configurations for four *equal* masses have been determined by Albouy (1995, 1996).

Exercises

6.1 Give a geometric interpretation of Proposition 6.10 in the case $p = 0$.

6.2 (a) Let $\Omega \subset \mathbb{R}^d$ be an open cone, and $f\colon \Omega \to \mathbb{R}$ a C^1-function satisfying

$$f(\lambda\mathbf{x}) = \log\lambda + f(\mathbf{x}) \quad \text{for all } \mathbf{x} \in \Omega \text{ and } \lambda \in \mathbb{R}^+.$$

Show that

$$\langle \operatorname{grad} f(\mathbf{x}), \mathbf{x} \rangle = 1 \quad \text{for all } \mathbf{x} \in \Omega.$$

(b) Show that the n-body problem for the force law $F \propto r^{-1}$ does not have any equilibrium solutions.

6.3 Let $(\mathbf{r}_1, \mathbf{r}_2, \mathbf{r}_3)\colon (\alpha, \infty) \to \mathbb{R}^{3\cdot 3} \setminus \Delta$ be a solution of the three-body problem. We are not assuming that the centre of mass stays fixed. Show that it is not possible for $\mathbf{r}_1(t)$ and $\mathbf{r}_2(t)$ to converge to the same point in $\mathbb{R}^3$ as $t \to \infty$.

Hint: Argue by contradiction. Study the behaviour of the moment of inertia $I(t)$ and of $\mathbf{r}_3(t)$ as $t \to \infty$ under the assumption

$$\lim_{t\to\infty} \mathbf{r}_1(t) = \mathbf{p} = \lim_{t\to\infty} \mathbf{r}_2(t).$$

6.4 How are the two sides of the Lagrange–Jacobi identity $\ddot{I} = 2T + V(\mathbf{r})$

affected when a solution $\mathbf{r}$ of the n-body problem is replaced by the solution $t \mapsto \mathbf{r}(t) + \mathbf{a}t + \mathbf{b}$?

6.5 Let $\mathbf{r} = (\mathbf{r}_1, \dots, \mathbf{r}_n)\colon (\alpha, \omega) \to \mathbb{R}^{3n} \setminus \Delta$ be a maximal solution of the n-body problem that is bounded and bounded away from collisions, i.e. there are constants $C, \varepsilon > 0$ such that $|\mathbf{r}_i(t)| \leq C$ and $|\mathbf{r}_i(t) - \mathbf{r}_j(t)| \geq \varepsilon$ for all $t \in (\alpha, \omega)$ and $i, j \in \{1, \dots, n\}$.

(a) Show that the velocities $\dot{\mathbf{r}}_i$ likewise stay bounded, and conclude that $(\alpha, \omega) = \mathbb{R}$.

(b) Show that the energy h of $\mathbf{r}$ is negative.

6.6 A subset $A \subset \mathbb{R}^d$ is called **convex** if $s\mathbf{x} + (1 - s)\mathbf{y} \in A$ for all $\mathbf{x}, \mathbf{y} \in A$ and $s \in [0, 1]$.

(a) Let $\{A_\lambda\}_{\lambda \in \Lambda}$ be a family of convex subsets of $\mathbb{R}^d$. Show that their intersection $\bigcap_{\lambda \in \Lambda} A_\lambda \subset \mathbb{R}^d$ is likewise convex.

(b) Let $S \subset \mathbb{R}^d$ be an arbitrary subset. The **convex hull** of S is

$$\mathrm{Co}(S) := \bigcap_{\substack{A \supset S \\ A \text{ convex}}} A.$$

Show that $\mathrm{Co}(S)$ is the smallest convex set containing S.

(c) Let points $\mathbf{p}_1, \dots, \mathbf{p}_n \in \mathbb{R}^d$ be given. Show that the convex hull of these points can be described as

$$\mathrm{Co}(\{\mathbf{p}_1, \dots, \mathbf{p}_n\}) = \left\{ \sum_{i=1}^n s_i \mathbf{p}_i \colon\ s_i \geq 0,\ \sum_{i=1}^n s_i = 1 \right\}.$$

6.7 Let $\mathbf{r}$ be a solution of the n-body problem with centre of mass $\mathbf{r}_c$ fixed at the origin. Compare its angular momentum with that of the solution $t \mapsto \mathbf{r}(t) + \mathbf{a}t + \mathbf{b}$.

6.8 Let $t \mapsto \mathbf{r}(t) \in \mathbb{R}^{3n} \setminus \Delta$ describe an arbitrary motion of n bodies in 3-space, $\mathbf{r} = (\mathbf{r}_1, \dots, \mathbf{r}_n)$, with centre of mass fixed at the origin. Set $r_{ij} := |\mathbf{r}_i - \mathbf{r}_j|$ and $M := m_1 + \dots + m_n$. Show that the moment of inertia I can be computed from these pairwise distances as

$$I = \frac{1}{4M} \sum_{i,j=1}^n m_i m_j r_{ij}^2.$$

6.9 Let $\mathbf{r} = (\mathbf{r}_1, \dots, \mathbf{r}_n)$ be a solution of the n-body problem with centre of mass fixed at the origin. We use the notation of the preceding exercise, and we set $v_{ij} := |\mathbf{v}_i - \mathbf{v}_j|$, where $\mathbf{v}_i = \dot{\mathbf{r}}_i$. Our aim in this exercise is to derive the inequality

$$c^2 + \dot{I}^2 \leq 4IT,$$

which is a stronger version of Sundman's inequality. Analogous to the foregoing exercise we have

$$T = \frac{1}{4M} \sum_{i,j=1}^{n} m_i m_j v_{ij}^2,$$

and as in Exercise 1.6 we have

$$|(\mathbf{r}_i - \mathbf{r}_j) \times (\mathbf{v}_i - \mathbf{v}_j)|^2 = r_{ij}^2 \, (v_{ij}^2 - \dot{r}_{ij}^2).$$

(a) Set

$$Q := \frac{1}{4M} \sum_{i,j=1}^{n} \dot{r}_{ij}^2.$$

Verify the inequality

$$\dot{I}^2 \leq 4IQ. \tag{6.5}$$

(b) Show that the angular momentum $\mathbf{c}$ can be written as

$$\mathbf{c} = \frac{1}{2M} \sum_{i,j=1}^{n} m_i m_j (\mathbf{r}_i - \mathbf{r}_j) \times (\mathbf{v}_i - \mathbf{v}_j).$$

(c) Combine (b) with the observations at the beginning of this exercise to arrive at the inequality

$$c^2 \leq 4I(T - Q). \tag{6.6}$$

The strong version of Sundman's inequality follows by summing (6.5) and (6.6).

6.10 The Sundman inequality $c^2 \leq 4IT$ followed from a simple application of the Cauchy–Schwarz inequality. For the stronger version $c^2 + \dot{I}^2 \leq 4IT$ in the previous exercise, you will also have used the triangle inequality. Check the case of equality in these arguments and derive the following statements for a solution of the n-body problem with centre of mass fixed at the origin.

(a) We have equality $c^2 + \dot{I}^2 = 4IT$ in the stronger version of Sundman's inequality if and only if the motion is planar homographic, and hence comes from a planar central configuration.

(b) Equality $c^2 = 4IT$ in the weaker version of Sundman's inequality is equivalent to the motion being a planar relative equilibrium rotating with constant angular velocity.

6.11 In the three-body problem with centre of mass fixed at the origin we consider the so-called **Jacobi coordinates**

$$\mathbf{r} := \mathbf{r}_2 - \mathbf{r}_1 \quad \text{and} \quad \mathbf{R} := Mm^{-1}\mathbf{r}_3,$$

where $m := m_1 + m_2$ and $M := m_1 + m_2 + m_3$. Show the following.

(a) The vector $\mathbf{R}$ describes the position of m_3 relative to the centre of mass of m_1 and m_2 (see the footnote on page 84).

(b) $\mathbf{r}_3 - \mathbf{r}_1 = \mathbf{R} + m_2 m^{-1}\mathbf{r}$ and $\mathbf{r}_3 - \mathbf{r}_2 = \mathbf{R} - m_1 m^{-1}\mathbf{r}$.

(c) The differential equations for the three-body problem can be written as

$$\ddot{\mathbf{r}} = -\frac{Gm}{r^3}\mathbf{r} + Gm_3\left(\frac{\mathbf{R} - m_1 m^{-1}\mathbf{r}}{r_{23}^3} - \frac{\mathbf{R} + m_2 m^{-1}\mathbf{r}}{r_{13}^3}\right),$$

$$\ddot{\mathbf{R}} = -\frac{GMm_1 m^{-1}}{r_{13}^3}(\mathbf{R} + m_2 m^{-1}\mathbf{r}) - \frac{GMm_2 m^{-1}}{r_{23}^3}(\mathbf{R} - m_1 m^{-1}\mathbf{r}),$$

where $r_{ij} := |\mathbf{r}_i - \mathbf{r}_j|$. Notice that the assumption on the centre of mass being fixed reduces the three-body problem to one of order 12.

(d) Set $\mathbf{v} := \dot{\mathbf{r}}$, $\mathbf{V} := \dot{\mathbf{R}}$, $a := m_1 m_2/m$ and $A := m_3 m/M$. Then

$$\begin{aligned} \mathbf{c} &= a(\mathbf{r} \times \mathbf{v}) + A(\mathbf{R} \times \mathbf{V}), \\ 2I &= ar^2 + AR^2, \\ 2T &= av^2 + AV^2. \end{aligned}$$

In other words, angular momentum, moment of inertia and kinetic energy look as in a two-body system with respective masses a, A, position vectors $\mathbf{r}, \mathbf{R}$, and velocities $\mathbf{v}, \mathbf{V}$.

6.12 We consider the case $c = 0$ in the three-body problem. Using Jacobi coordinates (see the foregoing exercise), we want to prove a theorem of Weierstraß according to which the condition $c = 0$ implies that the three bodies move in a fixed plane. In combination with Sundman's theorem this shows that a total collapse can occur only in *planar* three-body systems.

(a) Show that $\langle \mathbf{r}, \mathbf{R} \times \mathbf{V} \rangle = 0$ and conclude that $\mathbf{r}, \mathbf{R}, \mathbf{V}$ lie in a single plane. Show analogously that $\mathbf{r}, \mathbf{v}, \mathbf{R}$ lie in a single plane. Why may we not, in general, conclude that all four vectors lie in a single plane?

(b) Let us first consider the case that $\mathbf{r}(t)$ and $\mathbf{R}(t)$ are linearly independent for all t in some time interval. We want to show that $\mathbf{r}(t)$ and $\mathbf{R}(t)$ span the same plane for all t. To this end, verify the equation

$$(\mathbf{r} \times \mathbf{R}) \times \frac{\mathrm{d}}{\mathrm{d}t}(\mathbf{r} \times \mathbf{R}) = \mathbf{0},$$

and then use the formula for $\frac{d}{dt}(\mathbf{u}/u)$ from Section 1.1. Argue analogously for the pair $\mathbf{r}, \mathbf{v}$. Explain why $\mathbf{r}_1(t), \mathbf{r}_2(t), \mathbf{r}_3(t)$ have to lie in that same fixed plane.

(c) Now consider the case that for some time t_0 both $\mathbf{R}(t_0)$ and $\mathbf{v}(t_0)$ are a multiple of $\mathbf{r}(t_0)$. (Notice that $\mathbf{r}(t)$ is never the zero vector, whereas $\mathbf{R}(t)$ may well be zero.) Show the following statements.

(1) If $\mathbf{V}(t_0)$ is likewise a multiple of $\mathbf{r}(t_0)$, then $\mathbf{r}(t)$ and $\mathbf{R}(t)$ always lie on a fixed line. Argue further that $\mathbf{r}_1(t), \mathbf{r}_2(t), \mathbf{r}_3(t)$ must then lie on a fixed line.

(2) If $\mathbf{V}(t_0)$ is not a multiple of $\mathbf{r}(t_0)$, then $\frac{d}{dt}(\mathbf{r} \times \mathbf{R})|_{t=t_0} \neq \mathbf{0}$. Give a careful reasoning for why in this case, too, the motion of the $\mathbf{r}_i$ must be planar.

6.13 Show that the regular n-gon is a central configuration for n equal masses.

7

The three-body problem

> It is quite a three pipe problem, and I beg that you won't speak to me for fifty minutes.
>
> Arthur Conan Doyle, *The Red-Headed League*

We now specialise to the case of *three* bodies moving under mutual gravitational attraction. In Section 7.1 we give a detailed proof of Lagrange's theorem on homographic solutions to the three-body problem. Lagrange's argument includes, in principle, the collinear solutions found earlier by Euler. For simplicity, however, we shall prove Lagrange's theorem under the assumption that the triangle formed by the three bodies is at all times non-degenerate. Then, in Section 7.2, we give a direct derivation of Euler's solutions under some simplifying assumptions. In Section 7.3 we derive various general results about the so-called *restricted* three-body problem, where one of the three masses is assumed to be negligibly small compared with the other two.

7.1 Lagrange's homographic solutions

In Section 6.6 we have given a complete description of the non-collinear *planar* homographic solutions of the three-body problem. The following theorem of Lagrange says that all non-collinear homographic solutions of the three-body problem are in fact planar, and hence of the form described earlier.

Theorem 7.1 (Lagrange) *Let* $(\mathbf{r}_1, \mathbf{r}_2, \mathbf{r}_3)$ *be a non-collinear homographic solution of the three-body problem, with centre of mass fixed at the origin. Then the following statements hold.*

(i) *The plane determined by the three bodies remains fixed in space.*

(ii) *The sum of the Newtonian forces acting on each of the three bodies is directed towards the origin.*
(iii) *The three bodies form an equilateral triangle.*
(iv) *The three bodies move on conic sections that are similar to one another, each having the origin as one of its foci.*

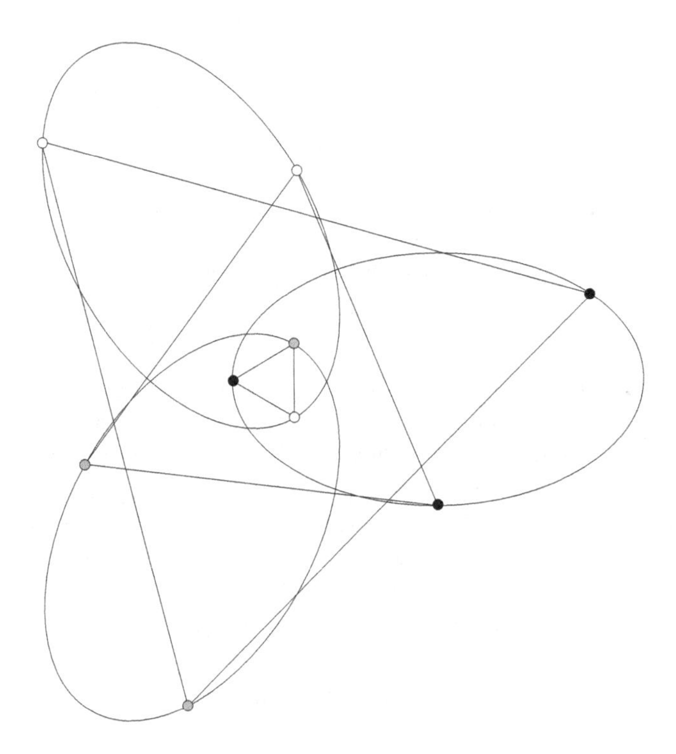

Figure 7.1 Lagrange's homographic solution for equal masses.

Figure 7.1 shows such a homographic solution in the case of three equal masses, where the three conic sections are congruent. Figure 7.2 shows a relative equilibrium with three different masses moving on circles.

Remark 7.2 Lagrange's solutions to the three-body problem can be observed in nature. In the plane of Jupiter's orbit around the Sun, near the two points forming an equilateral triangle with Sun and Jupiter, there are large groups of asteroids known as the *Trojans*. The ones staying ahead of Jupiter are named after Trojan heroes from Homer's *Iliad*; the ones chasing Jupiter, after Greek heroes – with the exception of the Greek Patroclus and the Trojan Hector, who have been placed in the wrong camp.

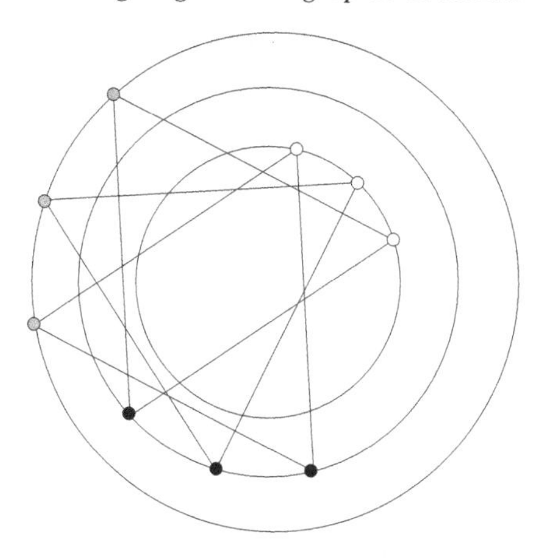

Figure 7.2 Lagrange's relative equilibrium solution.

Proof of Theorem 7.1 Once we have established that $(\mathbf{r}_1, \mathbf{r}_2, \mathbf{r}_3)$ is a planar motion, statements (ii) to (iv) are a direct consequence of Proposition 6.32 and Example 6.29, since the solution is homographic by assumption. So we only need to prove statement (i).

(1) We may write the solution $(\mathbf{r}_1, \mathbf{r}_2, \mathbf{r}_3)$ in the form

$$\mathbf{r}_i(t) = A(t)\mathbf{x}_i(t), \quad i = 1, 2, 3,$$

where $(\mathbf{x}_1, \mathbf{x}_2, \mathbf{x}_3)$ describe a motion in a fixed plane, and $A(t) \in \mathrm{SO}(3)$, the special orthogonal group. Without loss of generality we can take that plane to be the xy-plane, i.e.

$$\mathbf{x}_i(t) = \begin{pmatrix} a_i(t) \\ b_i(t) \\ 0 \end{pmatrix},$$

and we may assume $A(0) = E$, where E stands for the 3×3 unit matrix. For details, see Exercise 7.3.

(2) Before continuing with the problem at hand, we derive some general expressions regarding motions $t \mapsto \mathbf{x}(t)$ and $t \mapsto \mathbf{r}(t)$ related by

$$\mathbf{r} = A\mathbf{x} \tag{7.1}$$

for some SO(3)-valued function $t \mapsto A(t)$. The three functions $\mathbf{x}, \mathbf{r}, A$ are assumed to be of class at least C^2.

The condition $A \in \mathrm{SO}(3)$ means that $A^{\mathrm{t}}A = E$ (and $\det A = 1$). Hence

$$\dot{A}^{\mathrm{t}}A + A^{\mathrm{t}}\dot{A} = \mathbf{0},$$

i.e. $A^{\mathrm{t}}\dot{A}$ is skew-symmetric. (Observe that for any time-dependent matrix B we have

$$\left(\frac{\mathrm{d}}{\mathrm{d}t}B\right)^{\mathrm{t}} = \frac{\mathrm{d}}{\mathrm{d}t}(B^{\mathrm{t}}),$$

so the notation $\dot{B}^{\mathrm{t}}$ is unambiguous.) Write

$$A^{\mathrm{t}}\dot{A} = \begin{pmatrix} 0 & -\omega_3 & \omega_2 \\ \omega_3 & 0 & -\omega_1 \\ -\omega_2 & \omega_1 & 0 \end{pmatrix} \quad \text{and} \quad \omega = \begin{pmatrix} \omega_1 \\ \omega_2 \\ \omega_3 \end{pmatrix}.$$

Then

$$A^{\mathrm{t}}\dot{A}\mathbf{x} = \omega \times \mathbf{x}. \tag{7.2}$$

The vector ω is called the **instantaneous angular velocity**, for reasons explained in Exercise 7.4.

From (7.1) we have

$$\begin{aligned} \dot{\mathbf{r}} &= \dot{A}\mathbf{x} + A\dot{\mathbf{x}} \\ &= A(A^{\mathrm{t}}\dot{A}\mathbf{x} + \dot{\mathbf{x}}) \\ &= A(\omega \times \mathbf{x} + \dot{\mathbf{x}}) \end{aligned} \tag{7.3}$$

and

$$\ddot{\mathbf{r}} = \dot{A}(\omega \times \mathbf{x} + \dot{\mathbf{x}}) + A(\dot{\omega} \times \mathbf{x} + \omega \times \dot{\mathbf{x}} + \ddot{\mathbf{x}}).$$

With the help of (7.2) and the Graßmann identity (1.1), we can rewrite the latter in the more convenient form

$$\begin{aligned} A^{\mathrm{t}}\ddot{\mathbf{r}} &= \omega \times (\omega \times \mathbf{x} + \dot{\mathbf{x}}) + \dot{\omega} \times \mathbf{x} + \omega \times \dot{\mathbf{x}} + \ddot{\mathbf{x}} \\ &= \omega\langle\omega, \mathbf{x}\rangle - \mathbf{x}\langle\omega, \omega\rangle + 2\omega \times \dot{\mathbf{x}} + \dot{\omega} \times \mathbf{x} + \ddot{\mathbf{x}}. \end{aligned} \tag{7.4}$$

(3) According to the assumptions of the theorem, the solution $(\mathbf{r}_1, \mathbf{r}_2, \mathbf{r}_3)$ can be expressed as

$$\mathbf{r}_i(t) = A(t)\mathbf{x}_i(t)$$

with

$$\mathbf{x}_i(t) = \lambda(t)\begin{pmatrix} a_i \\ b_i \\ 0 \end{pmatrix}, \quad a_i, b_i \in \mathbb{R}, \quad \lambda(t) \in \mathbb{R}^+.$$

The assumption that the three bodies form a non-degenerate triangle is equivalent to

$$\operatorname{rank}\begin{pmatrix} a_1 & a_2 & a_3 \\ b_1 & b_2 & b_3 \end{pmatrix} = 2.$$

In fact, a little more is true. By assumption, the origin $\mathbf{0}$ is the centre of mass of the three bodies, so it has to lie in the interior of the triangle. Thus

$$\operatorname{rank}\begin{pmatrix} a_i & a_j \\ b_i & b_j \end{pmatrix} = 2 \quad \text{for} \quad i \neq j. \tag{7.5}$$

From $\mathbf{r}_j - \mathbf{r}_i = A(\mathbf{x}_j - \mathbf{x}_i)$ we have $|\mathbf{r}_j - \mathbf{r}_i| = |\mathbf{x}_j - \mathbf{x}_i|$, so the pairwise distances of the three bodies are proportional to λ. With the Newtonian equation

$$\ddot{\mathbf{r}}_i = \sum_{\substack{j=1 \\ j\neq i}}^{n} \frac{Gm_j}{|\mathbf{r}_j - \mathbf{r}_i|^2} \cdot \frac{\mathbf{r}_j - \mathbf{r}_i}{|\mathbf{r}_j - \mathbf{r}_i|}$$

we conclude further that

$$A^{\mathrm{t}}\, \ddot{\mathbf{r}}_i(t) = \frac{1}{\lambda^2(t)} \begin{pmatrix} L_i \\ M_i \\ 0 \end{pmatrix} \tag{7.6}$$

with some constants L_i, M_i.

(**4**) For better readability, write $\omega(t) = (p(t), q(t), r(t))^{\mathrm{t}}$. Then the three components of (7.4) take the following form:

$$\left.\begin{array}{rclcl} L_i/\lambda^2 &=& a_i(\ddot{\lambda} - (q^2 + r^2)\lambda) &-& b_i(2r\dot{\lambda} + (\dot{r} - pq)\lambda) \\ M_i/\lambda^2 &=& a_i(2r\dot{\lambda} + (\dot{r} + pq)\lambda) &+& b_i(\ddot{\lambda} - (p^2 + r^2)\lambda) \\ 0 &=& -a_i(2q\dot{\lambda} + (\dot{q} - pr)\lambda) &+& b_i(2p\dot{\lambda} + (\dot{p} + qr)\lambda). \end{array}\right\} \tag{7.7}$$

The first of these three equations for $i = 1, 2$, say, can be written as

$$\begin{pmatrix} L_1 \\ L_2 \end{pmatrix} = \begin{pmatrix} a_1 & b_1 \\ a_2 & b_2 \end{pmatrix} \cdot \begin{pmatrix} \lambda^2(\ddot{\lambda} - (q^2 + r^2)\lambda) \\ -\lambda^2(2r\dot{\lambda} + (\dot{r} - pq)\lambda) \end{pmatrix}.$$

The vector on the right-hand side of this equation does not depend on t by the rank condition (7.5). Likewise we argue with the second of the equations (7.7). Hence, by forming the crosswise difference of coefficients in the first two lines of (7.7) we obtain

$$p^2 - q^2 = \frac{a}{\lambda^3}, \quad 2pq = \frac{b}{\lambda^3} \tag{7.8}$$

with some constants a, b.

Similarly, applying the rank condition (7.5) to the last of the equations (7.7), we get

$$2q\dot{\lambda} + (\dot{q} - pr)\lambda = 0, \quad 2p\dot{\lambda} + (\dot{p} + qr)\lambda = 0. \tag{7.9}$$

This yields

$$\begin{aligned}0 &= q(2q\dot{\lambda} + (\dot{q} - pr)\lambda) + p(2p\dot{\lambda} + (\dot{p} + qr)\lambda)\\ &= 2(p^2 + q^2)\dot{\lambda} + (p\dot{p} + q\dot{q})\lambda\\ &= \frac{1}{2\lambda^3}\frac{\mathrm{d}}{\mathrm{d}t}((p^2 + q^2)\lambda^4),\end{aligned}$$

hence

$$p^2 + q^2 = \frac{c}{\lambda^4}$$

for some constant c.

On the other hand, the two equations (7.8) can be written in complex form as

$$(p \pm \mathrm{i}q)^2 = \frac{a \pm \mathrm{i}b}{\lambda^3},$$

hence

$$(p^2 + q^2)^2 = ((p + \mathrm{i}q)(p - \mathrm{i}q))^2 = \frac{(a + \mathrm{i}b)(a - \mathrm{i}b)}{\lambda^6} = \frac{a^2 + b^2}{\lambda^6}.$$

Combining this with the previous expression for $p^2 + q^2$, we find

$$\sqrt{a^2 + b^2}\,\lambda = c.$$

This gives two possibilities.

First case. Both $\sqrt{a^2 + b^2}$ and c are zero. In this case $p = q = 0$.

Second case. Both $\sqrt{a^2 + b^2}$ and c are non-zero. This forces λ to be constant. From the equations for p^2+q^2 and p^2-q^2 we see that these expressions are likewise constant, hence both p and q are constant (but not both equal to zero). Then from (7.9) we get $r = 0$, and hence from (7.7)

$$\frac{1}{\lambda^3}\begin{pmatrix}L_i\\ M_i\end{pmatrix} = \begin{pmatrix}-q^2 & pq\\ pq & -p^2\end{pmatrix}\begin{pmatrix}a_i\\ b_i\end{pmatrix}.$$

Since

$$\det\begin{pmatrix}-q^2 & pq\\ pq & -p^2\end{pmatrix} = 0,$$

this would imply by (7.6) that the forces acting on the three bodies are collinear, which requires the bodies themselves to be aligned.

So, under the non-degeneracy assumption of the theorem, only the first case can occur (but see Exercise 7.9).

(5) From $p = q = 0$, i.e. $\boldsymbol{\omega}$ being proportional to $(0, 0, 1)^{\mathrm{t}}$, it follows that the three bodies move in a fixed plane, which is statement (i) of the theorem.

Indeed, at time t the three bodies lie in the plane $A(t)(\mathbb{R}^2 \times \{0\})$. It therefore suffices to show that the vector $A(0,0,1)^{\mathrm{t}}$ remains constant, which is easy:

$$\frac{\mathrm{d}}{\mathrm{d}t}(A(0,0,1)^{\mathrm{t}}) = \dot{A}(0,0,1)^{\mathrm{t}} = AA^{\mathrm{t}}\dot{A}(0,0,1)^{\mathrm{t}} = A(\omega \times (0,0,1)^{\mathrm{t}}) = \mathbf{0},$$

where we have used (7.2). The assumption $A(0) = E$ then implies that the motion takes place in the xy-plane. □

7.2 Euler's collinear solutions

The argument we used to prove Theorem 7.1 can be adapted to the situation where the triangle formed by the three bodies is degenerate, that is, the three bodies are always collinear, see Exercise 7.9.

Here we give a direct derivation of such collinear solutions under two simplifying assumptions.

- The line determined by the three bodies stays in a fixed plane.
- The three bodies move on circular orbits.

Thus, we are looking for solutions of the three-body problem of the form

$$\mathbf{r}_j(t) = \mathrm{e}^{\mathrm{i}\omega t} x_j, \quad j = 1,2,3,$$

with $\omega \in \mathbb{R}^+$ and $x_j \in \mathbb{R} \subset \mathbb{C}$. Without loss of generality we may assume $x_1 < x_2 < x_3$.

The equations of the three-body problem then become

$$\left.\begin{aligned} -\omega^2 x_1 &= G\left(\frac{m_2}{(x_2-x_1)^2} + \frac{m_3}{(x_3-x_1)^2}\right) \\ -\omega^2 x_2 &= G\left(-\frac{m_1}{(x_2-x_1)^2} + \frac{m_3}{(x_3-x_2)^2}\right) \\ -\omega^2 x_3 &= G\left(-\frac{m_1}{(x_3-x_1)^2} - \frac{m_2}{(x_3-x_2)^2}\right). \end{aligned}\right\} \tag{7.10}$$

Multiplying the jth equation by m_j, $j = 1,2,3$, and taking the sum over the three equations yields $m_1x_1 + m_2x_2 + m_3x_3 = 0$, that is, for solutions of this type the centre of mass is automatically fixed at the origin. Set

$$a := x_2 - x_1 \quad \text{and} \quad \rho := \frac{x_3 - x_2}{x_2 - x_1},$$

so that

$$a\rho = x_3 - x_2 \quad \text{and} \quad a(1+\rho) = x_3 - x_1.$$

Then the difference between the first and the second of the equations (7.10) can be written as

$$\frac{\omega^2}{G}a = \frac{m_1 + m_2}{a^2} + \frac{m_3}{a^2(1+\rho)^2} - \frac{m_3}{a^2\rho^2}; \tag{7.11}$$

the difference between the second and the third equation becomes

$$\frac{\omega^2}{G}a\rho = \frac{m_2 + m_3}{a^2\rho^2} + \frac{m_1}{a^2(1+\rho)^2} - \frac{m_1}{a^2}. \tag{7.12}$$

Hence

$$m_1 + m_2 + \frac{m_3}{(1+\rho)^2} - \frac{m_3}{\rho^2} = \frac{\omega^2}{G}a^3 = \frac{m_2 + m_3}{\rho^3} + \frac{m_1}{\rho(1+\rho)^2} - \frac{m_1}{\rho}.$$

Multiplying by $\rho^3(1+\rho)^2$ and collecting powers of ρ gives

$$\begin{aligned}&(m_1 + m_2)\rho^5 + (3m_1 + 2m_2)\rho^4 + (3m_1 + m_2)\rho^3 \\ &- (m_2 + 3m_3)\rho^2 - (2m_2 + 3m_3)\rho - (m_2 + m_3) = 0.\end{aligned} \tag{7.13}$$

This polynomial equation has a unique solution ρ in $\mathbb{R}^+$, see Exercise 7.8. With (7.11) one then finds $\omega^2 a^3$; by the choice of ρ as a solution of (7.13), this is consistent with the value of $\omega^2 a^3$ one finds using (7.12). The value of $a \in \mathbb{R}^+$ may be chosen freely, this is simply a scaling factor. Then x_j is determined by the jth equation in (7.10).

7.3 The restricted three-body problem

In the *restricted* three-body problem one makes the simplifying assumption that the mass m_3 is negligibly small compared with m_1 and m_2. Formally, this means passing to the limit $m_3 \to 0$ in the equations of the three-body problem, which yields the following system:

$$\left.\begin{aligned}\ddot{\mathbf{r}}_1 &= \frac{Gm_2}{|\mathbf{r}_2 - \mathbf{r}_1|^2}\cdot\frac{\mathbf{r}_2 - \mathbf{r}_1}{|\mathbf{r}_2 - \mathbf{r}_1|}\\ \ddot{\mathbf{r}}_2 &= \frac{Gm_1}{|\mathbf{r}_1 - \mathbf{r}_2|^2}\cdot\frac{\mathbf{r}_1 - \mathbf{r}_2}{|\mathbf{r}_1 - \mathbf{r}_2|}\\ \ddot{\mathbf{r}}_3 &= \frac{Gm_1}{|\mathbf{r}_1 - \mathbf{r}_3|^2}\cdot\frac{\mathbf{r}_1 - \mathbf{r}_3}{|\mathbf{r}_1 - \mathbf{r}_3|} + \frac{Gm_2}{|\mathbf{r}_2 - \mathbf{r}_3|^2}\cdot\frac{\mathbf{r}_2 - \mathbf{r}_3}{|\mathbf{r}_2 - \mathbf{r}_3|}.\end{aligned}\right\} \tag{R3B}$$

The first two equations describe the two-body problem for[1] m_1 and m_2. Thus, by Chapter 5, we may assume that $\mathbf{r}_1$ and $\mathbf{r}_2$ are given as functions

[1] I use the symbol m_i also as shorthand for 'the body of mass m_i'.

of time. Then the third equation of (R3B) is an equation for $\mathbf{r}_3$ only. Notice, however, that this is then a so-called *non-autonomous* equation, i.e. the right-hand side depends explicitly on time via $\mathbf{r}_1(t)$ and $\mathbf{r}_2(t)$. In the case of practical interest, where m_1 and m_2 move on elliptic orbits, this time dependence is periodic.

We make two further simplifying assumptions.

- The body m_3 moves in the plane determined by the motion of the **primaries** m_1 and m_2.
- The primaries move on circular orbits around their common centre of mass, which is fixed at the origin.

With these additional assumptions, we speak of the *planar circular* restricted three-body problem (PCR3B).

Number the primaries such that $m_1 \geq m_2$, and choose $m_1 + m_2$ as the unit for mass, i.e. set $m_1 + m_2 = 1$. Define $\mu := m_2 \in (0, 1/2]$; then $m_1 = 1 - \mu$.

The circularity assumption allows us to write

$$\mathbf{r}_1(t) = \mathrm{e}^{\mathrm{i}\omega t}\mathbf{z}_1, \quad \mathbf{r}_2(t) = \mathrm{e}^{\mathrm{i}\omega t}\mathbf{z}_2$$

for some $\mathbf{z}_1, \mathbf{z}_2 \in \mathbb{C}$. The condition that the centre of mass be fixed at the origin becomes

$$(1 - \mu)\mathbf{z}_1 + \mu\mathbf{z}_2 = \mathbf{0}.$$

By a translation of time or a rotation of the coordinate system we may assume that $\mathbf{z}_1 \in \mathbb{R}^-$ and $\mathbf{z}_2 \in \mathbb{R}^+$. Choose the unit of length such that $\mathbf{z}_2 - \mathbf{z}_1 = 1$. Then

$$\boxed{\mathbf{z}_1 = (-\mu, 0), \quad \mathbf{z}_2 = (1 - \mu, 0).}$$

Finally, choose the unit of time such that $\omega = 1$, i.e. the motion of the primaries is periodic of period 2π. The first equation of (R3B) implies that in these units we have $G = 1$.

We now describe the motion of m_3 in a coordinate system that rotates with the primaries, i.e. we set

$$\mathbf{r}_3(t) = \mathrm{e}^{\mathrm{i}t}\mathbf{z}(t).$$

Then

$$\dot{\mathbf{r}}_3(t) = \mathrm{e}^{\mathrm{i}t}(\mathrm{i}\mathbf{z}(t) + \dot{\mathbf{z}}(t))$$

and

$$\ddot{\mathbf{r}}_3(t) = \mathrm{e}^{\mathrm{i}t}(-\mathbf{z} + 2\mathrm{i}\dot{\mathbf{z}} + \ddot{\mathbf{z}}).$$

Inserting this into the third equation of (R3B) gives

$$\ddot{\mathbf{z}} + 2\mathrm{i}\dot{\mathbf{z}} = \mathbf{z} - \frac{1-\mu}{|\mathbf{z}-\mathbf{z}_1|^3}(\mathbf{z}-\mathbf{z}_1) - \frac{\mu}{|\mathbf{z}-\mathbf{z}_2|^3}(\mathbf{z}-\mathbf{z}_2). \tag{7.14}$$

We are seeking a solution $\mathbf{z} \in C^2(\mathbb{R}, \mathbb{C} \setminus \{\mathbf{z}_1, \mathbf{z}_2\})$ of this differential equation. Notice that by working in a rotating coordinate system we have arrived at an *autonomous* equation – there is no longer an explicit time dependence. The price we have to pay is the appearance of two new terms, $2\mathrm{i}\dot{\mathbf{z}}$ and $\mathbf{z}$, that do not derive directly from the law of gravitation.

It is convenient to set

$$\boxed{\Phi(\mathbf{z}) := \frac{1}{2}|\mathbf{z}|^2 + \frac{1-\mu}{|\mathbf{z}-\mathbf{z}_1|} + \frac{\mu}{|\mathbf{z}-\mathbf{z}_2|} + \frac{1}{2}\mu(1-\mu), \quad \mathbf{z} \in \mathbb{R}^2 \setminus \{\mathbf{z}_1, \mathbf{z}_2\},}$$

where the constant term $\mu(1-\mu)/2$ is a historical artefact, see also Exercise 7.11. We have reverted to interpreting $\mathbf{z}$ as a point in $\mathbb{R}^2$ rather than $\mathbb{C}$. In this real notation, and with $J := \left(\begin{smallmatrix} 0 & -1 \\ 1 & 0 \end{smallmatrix}\right)$, the differential equation for $\mathbf{z}$ can be written in the form

$$\ddot{\mathbf{z}} + 2J\dot{\mathbf{z}} = \operatorname{grad} \Phi(\mathbf{z}), \tag{PCR3B}$$

and we are seeking a solution $\mathbf{z} \in C^2(\mathbb{R}, \mathbb{R}^2 \setminus \{\mathbf{z}_1, \mathbf{z}_2\})$. Notice that the matrix J describes a rotation through a right angle in the counter-clockwise direction, just like multiplication by i in the complex plane. The term $-2J\dot{\mathbf{z}}$ in this expression for the acceleration $\ddot{\mathbf{z}}$ can be interpreted as a Coriolis force.

7.3.1 The Jacobi constant

The assumptions of the planar circular restricted three-body problem may appear to be quite restrictive, but the problem is not without physical relevance. For instance, the eccentricity of Jupiter's orbit around the Sun is about 0.05, so a circular orbit is a reasonable approximation. The masses of the Trojan asteroids mentioned in Remark 7.2 are negligible compared with those of the Sun and Jupiter. Hence, (PCR3B) is a good model for the Trojans. Indeed, we shall recover the Lagrange solutions as equilibrium points in (PCR3B).

When the motion of the primaries is taken as given, (R3B) is of order 6, as a second-order equation in three spatial variables. Because of the physically unrealistic assumption $m_3 = 0$ the usual conservation laws do not hold, so the order cannot be reduced. Under the assumption of planarity the order is reduced to 4; in the circular case, this is evident from the explicit equation (PCR3B). As we shall see presently, here the order can in fact be reduced to 3.

With $\mathbf{z} = (x, y)^{\mathsf{t}}$, the equation (PCR3B) can be rewritten as

$$\left.\begin{aligned} \ddot{x} - 2\dot{y} &= \frac{\partial \Phi}{\partial x} \\ \ddot{y} + 2\dot{x} &= \frac{\partial \Phi}{\partial y}. \end{aligned}\right\}$$

Multiplying the first equation in this system by $\dot{x}$ and the second by $\dot{y}$, and then summing the two equations, we obtain

$$\dot{x}\ddot{x} + \dot{y}\ddot{y} = \frac{\partial \Phi}{\partial x}\dot{x} + \frac{\partial \Phi}{\partial y}\dot{y}.$$

Integration gives

$$\dot{x}^2 + \dot{y}^2 = 2\Phi - C,$$

where C is a constant of integration (and the minus sign is another historical artefact). This last equation gives us one further constant of motion, as promised. The value C for a given solution of (PCR3B) is called the **Jacobi constant** of that solution.

I now wish to show how such a constant of motion can be interpreted geometrically. Write (PCR3B) as a system of first-order equations on the open subset $\Omega := (\mathbb{R}^2 \setminus \{\mathbf{z}_1, \mathbf{z}_2\}) \times \mathbb{R}^2$ of $\mathbb{R}^4$:

$$\left.\begin{aligned} \dot{\mathbf{z}} &= \mathbf{w} \\ \dot{\mathbf{w}} &= -2J\mathbf{w} + \operatorname{grad} \Phi(\mathbf{z}). \end{aligned}\right\} \qquad \text{(PCR3B}')$$

Definition 7.3 The function $\mathcal{J}$ on Ω, defined by

$$\mathcal{J}(\mathbf{z}, \mathbf{w}) := 2\Phi(\mathbf{z}) - |\mathbf{w}|^2,$$

is called the **Jacobi integral**.

Lemma 7.4 *If $t \mapsto \mathbf{z}(t)$ is a solution of* (PCR3B)*, then $\mathcal{J}(\mathbf{z}, \dot{\mathbf{z}})$ is constant.*

Proof This is exactly what our considerations above have shown; $\mathcal{J}(\mathbf{z}, \dot{\mathbf{z}})$ is the Jacobi constant of $\mathbf{z}$. However, we can also compute more succinctly:

$$\begin{aligned} \frac{\mathrm{d}(\mathcal{J}(\mathbf{z}, \dot{\mathbf{z}}))}{\mathrm{d}t} &= \langle 2 \operatorname{grad} \Phi(\mathbf{z}), \dot{\mathbf{z}} \rangle - 2\langle \ddot{\mathbf{z}}, \dot{\mathbf{z}} \rangle \\ &= 4\langle J\dot{\mathbf{z}}, \dot{\mathbf{z}} \rangle, \end{aligned}$$

which equals zero, since $J\dot{\mathbf{z}}$ is orthogonal to $\dot{\mathbf{z}}$. □

Remark 7.5 Notice the formal similarity, up to a factor -2, of

$$\mathcal{J}(\mathbf{z}, \dot{\mathbf{z}}) = 2\Phi(\mathbf{z}) - |\dot{\mathbf{z}}|^2$$

with the total energy as in Proposition 6.3. In the Hamiltonian formulation of (PCR3B) in Chapter 9, $-\mathcal{J}/2$ will be the Hamiltonian function, see Remark 9.1.

One ought to expect that prescribing a value for a constant of motion such as the Jacobi integral $\mathcal{J}$ reduces by one the number of degrees of freedom of the system in question. Indeed, if $C \in \mathbb{R}$ is chosen such that the gradient

$$\operatorname{grad} \mathcal{J}(\mathbf{z}, \mathbf{w}) = (2 \operatorname{grad} \Phi(\mathbf{z}), -2\mathbf{w})$$

does not vanish on the level set $M := \mathcal{J}^{-1}(C)$, then M is a three-dimensional submanifold of Ω (or the empty set) by the implicit function theorem.

With $\mathbf{x} := (\mathbf{z}, \mathbf{w})$ and

$$\mathbf{X}(\mathbf{z}, \mathbf{w}) := (\mathbf{w}, -2J\mathbf{w} + \operatorname{grad} \Phi(\mathbf{z})),$$

the system (PCR3B′) can be written as

$$\dot{\mathbf{x}} = \mathbf{X}(\mathbf{x}),$$

i.e. the solutions of the system are simply the integral curves of the vector field $\mathbf{X}$. An alternative expression of Lemma 7.4 is that this vector field is orthogonal to $\operatorname{grad} \mathcal{J}$, and hence tangent to the level set M:

$$\begin{aligned} \langle \operatorname{grad} \mathcal{J}(\mathbf{z}, \mathbf{w}), \mathbf{X}(\mathbf{z}, \mathbf{w}) \rangle_{\mathbb{R}^4} &= \langle 2 \operatorname{grad} \Phi(\mathbf{z}), \mathbf{w} \rangle_{\mathbb{R}^2} \\ &\quad + \langle -2\mathbf{w}, -2J\mathbf{w} + \operatorname{grad} \Phi(\mathbf{z}) \rangle_{\mathbb{R}^2} \\ &= 0, \end{aligned}$$

where we have used the fact that $J\mathbf{w}$ is orthogonal to $\mathbf{w}$.

This means that the integral curves $\mathbf{x}$ of $\mathbf{X}$ starting on M actually stay in M:

$$\begin{aligned} \frac{\mathrm{d}}{\mathrm{d}t} \mathcal{J}(\mathbf{x}(t)) &= \langle \operatorname{grad} \mathcal{J}(\mathbf{x}(t)), \dot{\mathbf{x}}(t) \rangle \\ &= \langle \operatorname{grad} \mathcal{J}(\mathbf{x}(t)), \mathbf{X}(\mathbf{x}(t)) \rangle \\ &= 0. \end{aligned}$$

In Chapter 10 we shall return to this topological point of view in the context of the Kepler problem.

7.3.2 The five libration points

We now wish to determine equilibrium solutions of (PCR3B), i.e. solutions with $\mathbf{z} = \text{const.}$, where m_3 stays fixed relative to the rotating coordinate system. This happens exactly at the points $\mathbf{z} \in \mathbb{R}^2 \setminus \{\mathbf{z}_1, \mathbf{z}_2\}$ where $\operatorname{grad} \Phi(\mathbf{z}) = \mathbf{0}$, that is,

$$\mathbf{z} - \frac{1-\mu}{\rho_1^3}(\mathbf{z} - \mathbf{z}_1) - \frac{\mu}{\rho_2^3}(\mathbf{z} - \mathbf{z}_2) = \mathbf{0}, \tag{7.15}$$

where $\rho_i := |\mathbf{z} - \mathbf{z}_i|$, $i = 1, 2$; see also equation (7.14).

First case. $y = 0$, and hence $x \in \mathbb{R} \setminus \{-\mu, 1-\mu\}$.
With $\mathbf{z} = (x, 0)$ we have $\mathbf{z} - \mathbf{z}_1 = (x + \mu, 0)$ and $\mathbf{z} - \mathbf{z}_2 = (x + \mu - 1, 0)$. So we are seeking solutions of the equation

$$x - (1-\mu)\frac{x+\mu}{|x+\mu|^3} - \mu\frac{x+\mu-1}{|x+\mu-1|^3} = 0. \tag{7.16}$$

(1a) $-\mu < x < 1-\mu$.
In this case, equation (7.16) becomes

$$f(x) := x - \frac{1-\mu}{(x+\mu)^2} + \frac{\mu}{(x+\mu-1)^2} = 0.$$

We have

$$f'(x) = 1 + \frac{2(1-\mu)}{(x+\mu)^3} - \frac{2\mu}{(x+\mu-1)^3} > 0 \text{ on } (-\mu, 1-\mu)$$

and

$$\lim_{x \searrow -\mu} f(x) = -\infty, \quad \lim_{x \nearrow 1-\mu} f(x) = \infty.$$

It follows that there is a unique $x_1 \in (-\mu, 1-\mu)$ with $f(x_1) = 0$. Write $L_1 := (x_1, 0)$ for the corresponding equilibrium point.

Observe that for $\mu = 1/2$ we have $f(0) = 0$, hence $x_1 = 0$; that is, L_1 is then the centre of mass of the primaries. For $\mu < 1/2$ we have

$$f(0) = -\frac{1-\mu}{\mu^2} + \frac{\mu}{(1-\mu)^2} < 0,$$

which implies $x_1 > 0$, so in this case L_1 lies between the centre of mass and the body of smaller mass.

(1b) $x > 1-\mu$.
Now (7.16) can be written as

$$f(x) := x - \frac{1-\mu}{(x+\mu)^2} - \frac{\mu}{(x+\mu-1)^2} = 0.$$

From

$$f'(x) = 1 + \frac{2(1-\mu)}{(x+\mu)^3} + \frac{2\mu}{(x+\mu-1)^3} > 0 \text{ on } (1-\mu, \infty)$$

and

$$\lim_{x \searrow 1-\mu} f(x) = -\infty, \quad \lim_{x \to \infty} f(x) = \infty$$

we conclude that there is a unique $x_2 \in (1-\mu, \infty)$ with $f(x_2) = 0$. Set $L_2 := (x_2, 0)$.

(1c) $x < -\mu$.
Here (7.16) becomes

$$f(x) := x + \frac{1-\mu}{(x+\mu)^2} + \frac{\mu}{(x+\mu-1)^2} = 0.$$

We see that

$$f'(x) = 1 - \frac{2(1-\mu)}{(x+\mu)^3} - \frac{2\mu}{(x+\mu-1)^3} > 0 \text{ on } (-\infty, -\mu),$$

and

$$\lim_{x\to-\infty} f(x) = -\infty, \quad \lim_{x\nearrow-\mu} f(x) = \infty.$$

Once again we conclude that there is a unique $x_3 \in (-\infty, -\mu)$ where $f(x_3) = 0$, and we set $L_3 := (x_3, 0)$.

Second case. $y \neq 0$.
In this case we consider the x- and y-components of (7.15) separately:

$$\left.\begin{aligned} x\left(1 - \frac{1-\mu}{\rho_1^3} - \frac{\mu}{\rho_2^3}\right) + \mu(1-\mu)\left(\frac{1}{\rho_2^3} - \frac{1}{\rho_1^3}\right) &= 0 \\ 1 - \frac{1-\mu}{\rho_1^3} - \frac{\mu}{\rho_2^3} &= 0. \end{aligned}\right\}$$

Combining the two equations we find that $\rho_1 = \rho_2 =: \rho$. Then, from the second equation, we conclude $\rho = 1$.

Our choice of units was such that $|\mathbf{z}_1 - \mathbf{z}_2| = 1$. So we see that in this second case we obtain two further equilibrium points L_4, L_5, characterised as the two points that form an equilateral triangle with $\mathbf{z}_1$ and $\mathbf{z}_2$.

In summary, we see that the only equilibrium solutions in (PCR3B) are the Lagrange and the Euler solutions described previously in greater generality; see also Exercise 7.12.

Definition 7.6 The points $L_1, \ldots, L_5$ are called the **libration points** in the planar circular restricted three-body problem, see Figure 7.3. The three points L_1, L_2, L_3 are the **Euler points**; L_4 and L_5 are referred to as the **Lagrange points**.

7.3.3 Hill's regions

The Jacobi constant of a solution $\mathbf{z}$ to (PCR3B) is $C = 2\Phi(\mathbf{z}) - |\dot{\mathbf{z}}|^2$. This places the constraint $2\Phi(\mathbf{z}) - C \geq 0$ on the possible positions of the third body. Conversely, given any point satisfying this constraint, one can find a

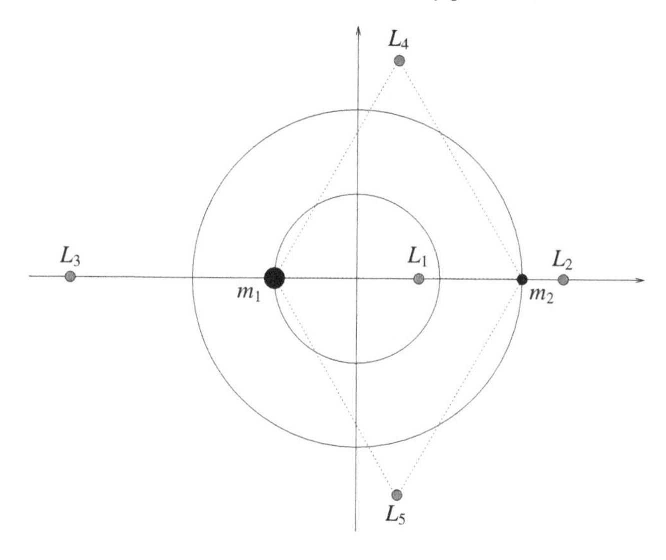

Figure 7.3 The five libration points in (PCR3B).

solution through that point with the given value of the Jacobi constant. This is analogous to the situation in the Kepler problem, where the motions for a given energy $h < 0$ are confined to a disc of radius $2a = -\mu/h$ about the origin, see Exercise 3.7.

Definition 7.7 **Hill's region** for the Jacobi constant C is the set

$$\{\mathbf{z} \in \mathbb{R}^2 \setminus \{\mathbf{z}_1, \mathbf{z}_2\} : \ 2\Phi(\mathbf{z}) - C \geq 0\}.$$

The curves forming its boundary, i.e. the set where the equality $2\Phi(\mathbf{z}) = C$ holds, are called the **zero velocity curves**.

Figure 7.4 shows some level curves of Φ. In Exercise 7.11 you are asked to verify that the function Φ attains its minimal value $\Phi = 3/2$ at the two Lagrange points L_4, L_5. This means that, for $C \leq 3$, Hill's region equals all of $\mathbb{R}^2 \setminus \{\mathbf{z}_1, \mathbf{z}_2\}$. For C slightly above the minimal value 3 of 2Φ, there are two roughly circular zero velocity curves about the two Lagrange points, and the motion is confined to the exterior of these circles. As the value of C increases a little further, the zero velocity curves become crescent-shaped, see the two dotted curves in Figure 7.4.

Observe that $\Phi(\mathbf{z})$ goes to infinity for $\mathbf{z} \to \mathbf{z}_1$ or $\mathbf{z} \to \mathbf{z}_2$. On the other hand, if $|\mathbf{z}|$ becomes large, Φ is essentially of the form $|\mathbf{z}|^2/2$. This means that, for large values of C, there are three circular zero velocity curves: two small ones about the primaries and a large one encircling both primaries. An example is given by the solid curves in Figure 7.4. The motion is confined to the interior of one of the small circular curves, or the exterior of the large one.

By computing the Hessian of Φ, one can show that the Euler points L_1, L_2, L_3 are saddle points of Φ. Moreover, it can be shown[2] that $\Phi(L_3) \leq \Phi(L_2) \leq \Phi(L_1)$.

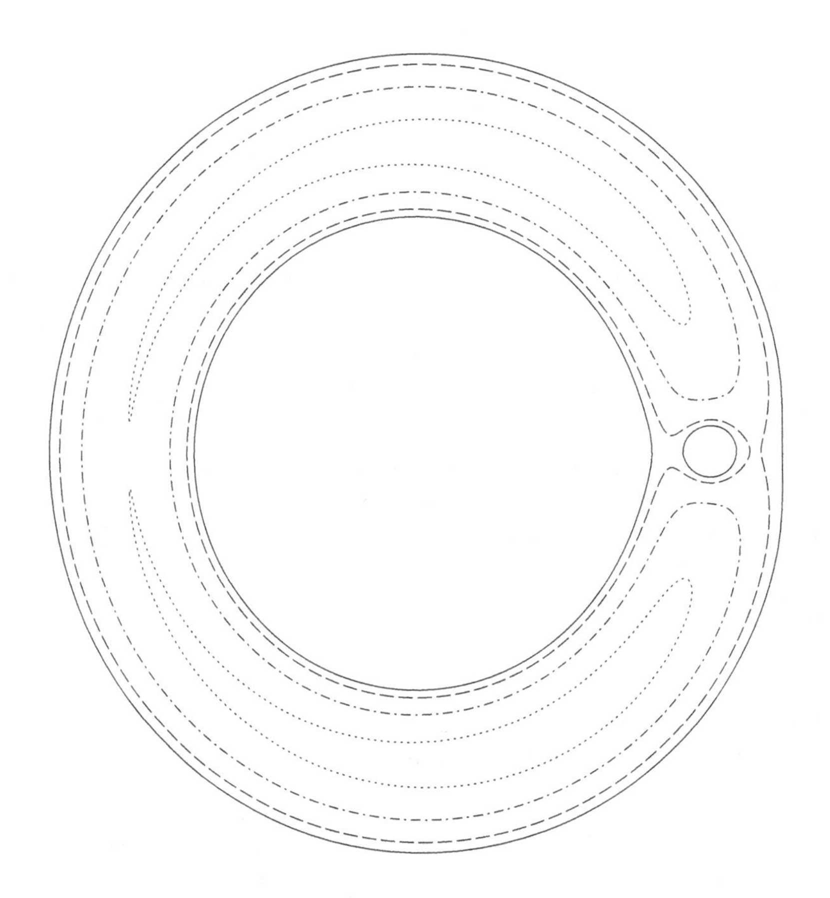

Figure 7.4 Level sets of the potential function Φ.

Notice that the topology of Hill's region changes whenever C passes the value of 2Φ at one of the libration points; see also Exercise 7.11. For instance, as the value of C passes $\Phi(L_3)$, the two crescent-shaped curves merge at the point L_3 into a single curve.

[2] These elementary but lengthy computations can be found in Section 4.6 of (Szebehely, 1967).

Notes and references

Theorem 7.1 was proved by Lagrange (1772) in his celebrated monograph on the three-body problem. This solution of the three-body problem by elementary functions remains the only one of its kind. Notice that in Lagrange's theorem the solution is not, to start with, assumed to be planar. As we have seen, establishing planarity (i.e. statement (i)) is really the hardest part of the proof. A simpler proof of Lagrange's theorem was given by Laplace (1799) in Book X, Chapter VI of the second part of his five-volume *Mécanique Céleste*, published between 1799 and 1825; this proof, however, made planarity an *a priori* assumption.

The proof of Theorem 7.1 given here follows a paper by Carathéodory (1933). I have used some modifications of that proof suggested by Sommerfeld (1942); various minor amplifications are my own. The final part of the argument has been simplified by appealing to the general theory of homographic solutions and central configurations.

Lagrange believed his solutions to be of merely theoretical significance. "Cette recherche n'est à la verité de pure curiosité. [...] Nous allons examiner dans ce Chapitre quelques cas particuliers, où le Problème des trois corps se simplifie beaucoup et admet une solution exacte ou presque exacte; quoique ces cas n'aient pas lieu dans le Système du monde, nous croyons cependent qu'ils méritent l'attention des Géomètres." Observation has proved Lagrange wrong: as mentioned in Remark 7.2, the homographic solutions of Lagrange are indeed realised in our solar system. The first Trojan asteroid, now named *588 Achilles*, was discovered in 1906 by Max Wolf at the Lagrange point L_4 of the Sun–Jupiter system.

A remarkable feature of Lagrange's solution is that the three bodies move on fixed trajectories in a single plane. It took until the year 2000 to discover a further solution of the three-body problem that shares this feature. Chenciner and Montgomery (2000) found a solution in the case of three equal masses, where the three bodies move along a single figure-eight curve in a plane. The beautiful paper (Montgomery, 2015) gives an elementary introduction to the geometric ideas behind the discovery of this figure-eight solution.

Solutions of the n-body problem where all bodies move along a single curve are referred to as **choreographic**. The paper by Chenciner and Montgomery has given the quest for periodic orbits in celestial mechanics a tremendous boost. Hundreds of choreographic solutions (with varying numbers of bodies) have now been found, at least numerically, see for instance the web page

`http://www.maths.manchester.ac.uk/~jm/Choreographies/`

of James Montaldi.

Leonhard Euler wrote a number of papers concerned with the three-body problem. Although the libration points L_1, L_2, L_3 are attributed to him, it seems that only L_1 and L_2 can be found implicitly in his paper *Considerationes de motu corporum coelestium* (E304 in the Eneström Index of Euler's *Opera Omnia*), published in 1766. The paper *De motu rectilineo trium corporum se mutuo attrahentium* (E327, published 1767) deals with three bodies on a *fixed* line only. Both papers can be found in (Euler, 1960). The paper *Considérations sur le problème des trois corps* (E400, 1770) with a similar content has the advantage of being written in a living language, and it contains some additional material on the n-body problem. It is scheduled to appear in vol. II/26 of the *Opera Omnia*.

The Jacobi integral was introduced by Jacobi (1836). This short note can be found in volume IV of his collected works. In the variational formulation of classical mechanics (of which more in Chapter 9), there is a general principle, known as Noether's theorem, describing how symmetries of the Lagrangian function give rise to constants of motion. For an interpretation of the Jacobi integral in the light of this principle see (Giordano and Plastino, 1998).

For an appreciation of Poincaré's contributions to the three-body problem see (Barrow-Green, 1997) and (Chenciner, 2015).

Exercises

7.1 Let three bodies of respective masses m_1, m_2, m_3 be given. We want to find a general formula for the vertices $\mathbf{z}_1, \mathbf{z}_2, \mathbf{z}_2$ of an equilateral triangle in the plane such that the centre of mass of the three bodies, with m_j placed at the point $\mathbf{z}_j$, lies at the origin. It is convenient to think of the $\mathbf{z}_j$ as points in the complex plane $\mathbb{C}$.

(a) Show that the general description of three points $\mathbf{z}_1, \mathbf{z}_2, \mathbf{z}_3$ forming the vertices of an equilateral triangle is given by

$$\mathbf{z}_1 = \mathbf{z}_0 + s\mathrm{e}^{\mathrm{i}\theta}, \quad \mathbf{z}_2 = \mathbf{z}_0 + s\zeta\mathrm{e}^{\mathrm{i}\theta}, \quad \mathbf{z}_3 = \mathbf{z}_0 + s\overline{\zeta}\mathrm{e}^{\mathrm{i}\theta},$$

where $\mathbf{z}_0 \in \mathbb{C}$, $\theta \in \mathbb{R}$, $s \in \mathbb{R}^+$ and $\zeta := \mathrm{e}^{2\pi\mathrm{i}/3}$.

(b) Describe the general formula for $\mathbf{z}_1, \mathbf{z}_2, \mathbf{z}_3$ if one imposes the centre of mass condition $m_1\mathbf{z}_1 + m_2\mathbf{z}_2 + m_3\mathbf{z}_3 = \mathbf{0}$.

7.2 In this exercise we wish to derive a special case of Lagrange's theorem. To this end, we consider a motion $t \mapsto \mathbf{r}_j(t) \subset \mathbb{C}$, $j = 1, 2, 3$, of three bodies of respective masses m_j in the complex plane. We assume that the three bodies move with equal period on circular orbits, i.e.

$$\mathbf{r}_j(t) = \mathrm{e}^{\mathrm{i}\omega t}\mathbf{z}_j, \quad j = 1, 2, 3,$$

where $\mathbf{z}_1, \mathbf{z}_2, \mathbf{z}_3 \in \mathbb{C}$ are three points not on a single line, and $\omega \in \mathbb{R}^+$. We want to show that this circular ansatz yields a solution of the three-body problem if and only if the following three conditions are satisfied.

(i) The centre of mass lies at the origin, i.e. $m_1\mathbf{z}_1 + m_2\mathbf{z}_2 + m_3\mathbf{z}_3 = \mathbf{0}$.
(ii) The three bodies form an equilateral triangle of side length s, say.
(iii) We have $\omega = \sqrt{GM/s^3}$, where $M := m_1 + m_2 + m_3$.

Proceed as follows. Let us first assume that the $\mathbf{r}_j$ do indeed give a solution of the three-body problem.

(a) Formulate the equations of the three-body problem as equations in terms of the $\mathbf{z}_j$. Observe that $|\mathbf{r}_j(t) - \mathbf{r}_k(t)| = |\mathbf{z}_j - \mathbf{z}_k| =: r_{jk}$.
(b) Derive condition (i) from a suitable combination of these equations.
(c) Convince yourself that one of the three equations for the three-body problem, say the one coming from $\ddot{\mathbf{r}}_3$, can be replaced by equation (i), and that the system of equations can then be written in the form $A\mathbf{Z} = \mathbf{0}$, where $\mathbf{Z} := (\mathbf{z}_1, \mathbf{z}_2, \mathbf{z}_3)^{\mathrm{t}}$ and

$$A := \begin{pmatrix} \omega^2 - \dfrac{Gm_2}{r_{12}^3} - \dfrac{Gm_3}{r_{13}^3} & \dfrac{Gm_2}{r_{12}^3} & \dfrac{Gm_3}{r_{13}^3} \\ \dfrac{Gm_1}{r_{12}^3} & \omega^2 - \dfrac{Gm_1}{r_{12}^3} - \dfrac{Gm_3}{r_{23}^3} & \dfrac{Gm_3}{r_{23}^3} \\ m_1 & m_2 & m_3 \end{pmatrix}.$$

(d) Explain why the condition that $\mathbf{z}_1, \mathbf{z}_2, \mathbf{z}_3$ not be collinear implies that the rank of A equals 1. Deduce that $r_{12} = r_{23} = r_{13} =: s$, i.e. (ii).
(e) Conclude that $\det A = 0$, and use this to prove (iii).

Conversely, verify that a motion of the form $t \mapsto \mathrm{e}^{\mathrm{i}\omega t}\mathbf{z}_j$, subject to conditions (i) to (iii), solves the equations of the three-body problem.

7.3 Let $t \mapsto (\mathbf{r}_1(t), \mathbf{r}_2(t), \mathbf{r}_3(t))$, $t \in \mathbb{R}$, describe the motion of three bodies in $\mathbb{R}^3$, with their centre of mass fixed at the origin. It is assumed that the three points $\mathbf{r}_1(t), \mathbf{r}_2(t), \mathbf{r}_3(t)$ always form a non-degenerate triangle.

(a) Show that, at any time t, the vectors $\mathbf{r}_1(t), \mathbf{r}_2(t), \mathbf{r}_3(t)$ are pairwise linearly independent, and the plane determined by these three points contains the origin.
(b) Choose cartesian coordinates in $\mathbb{R}^3$ such that $\mathbf{r}_1(0), \mathbf{r}_2(0), \mathbf{r}_3(0)$ lie in the xy-plane. (One could rotate the coordinate system such that $\mathbf{r}_1(0)$ points in the positive x-direction, or even scale the coordinates

such that $\mathbf{r}_1(0)$ has unit length, but this does not simplify any of the arguments in the proof of Theorem 7.1.) Set

$$\mathbf{f}_1(t) := \mathbf{r}_1(t)/r_1(t)$$
$$\mathbf{f}_2(t) := \frac{\mathbf{r}_2(t) - \langle \mathbf{r}_2(t), \mathbf{r}_1(t)\rangle \cdot \mathbf{r}_1(t)}{|\mathbf{r}_2(t) - \langle \mathbf{r}_2(t), \mathbf{r}_1(t)\rangle \cdot \mathbf{r}_1(t)|}$$
$$\mathbf{f}_3(t) := \mathbf{f}_1(t) \times \mathbf{f}_2(t)$$

and $\mathbf{e}_i := \mathbf{f}_i(0)$, $i = 1, 2, 3$. Define

$$A(t) := \begin{pmatrix} | & | & | \\ \mathbf{f}_1(t) & \mathbf{f}_2(t) & \mathbf{f}_3(t) \\ | & | & | \end{pmatrix} \begin{pmatrix} | & | & | \\ \mathbf{e}_1 & \mathbf{e}_2 & \mathbf{e}_3 \\ | & | & | \end{pmatrix}^{-1}.$$

Explain why $A(t) \in SO(3)$ for all t. Obviously, $A(0)$ is the unit matrix. If the $\mathbf{r}_i$ are of class C^k, show that A is likewise of class C^k.

(c) Set $\mathbf{x}_i(t) := A(t)^{\mathrm{t}}\mathbf{r}_i(t)$. Show that $t \mapsto (\mathbf{x}_1(t), \mathbf{x}_2(t), \mathbf{x}_3(t))$ describes a motion in the xy-plane.

7.4 (a) Associated with a map $t \mapsto A(t) \in SO(3)$ we have the instantaneous angular velocity $\omega_A(t) \in \mathbb{R}^3$, characterised by

$$A^{\mathrm{t}}\dot{A}\mathbf{x} = \omega_A \times \mathbf{x}$$

for $\mathbf{x} \in \mathbb{R}^3$. Let $t \mapsto B(t)$ be another map with values in SO(3). Show that

$$\omega_{AB}(t) = B^{\mathrm{t}}(t)\omega_A(t) + \omega_B(t).$$

(b) A motion of three bodies in $\mathbb{R}^3$ that stay on a line through the origin at all times can be described in the form

$$\mathbf{r}_i(t) = A(t)\mathbf{x}_i(t), \quad i = 1, 2, 3,$$

with $\mathbf{x}_i(t) = (x_i(t), 0, 0)^{\mathrm{t}}$. The $\mathbf{r}_i$ do not change if we replace $A(t)$ by $A(t)B(t)$ with

$$B(t) = \begin{pmatrix} 1 & 0 & 0 \\ 0 & \cos\varphi(t) & -\sin\varphi(t) \\ 0 & \sin\varphi(t) & \cos\varphi(t) \end{pmatrix}.$$

Compute ω_B and show that $B(t)$ can be chosen in such a way that $\omega_{AB}(t)$ is of the form $(0, q(t), r(t))^{\mathrm{t}}$.

(c) Let $C \in SO(3)$ be a matrix not depending on time. Show that $\omega_{CA} = \omega_A$. As in Exercise 1.1 one sees that if $(\mathbf{r}_1, \mathbf{r}_2, \mathbf{r}_3)$ is a solution of the three-body problem, then so is $(C\mathbf{r}_1, C\mathbf{r}_2, C\mathbf{r}_3)$. So we may assume that $A(0)$ is the identity matrix without changing the instantaneous angular velocity. Show that, under this assumption, we have $\dot{A}(0)\mathbf{x} = \omega_A(0) \times \mathbf{x}$. This justifies the terminology 'instantaneous angular velocity' for the vector $\omega_A(0)$.

7.5 Verify the computations leading to equation (7.7).

7.6 In this exercise we wish to show that a homographic solution (Definition 6.31) is homothetic (in the sense described at the beginning of Section 6.6) if and only if the angular momentum of the system is zero.

Clearly, a homothetic solution has zero angular momentum. Conversely, let $(\mathbf{r}_1, \ldots, \mathbf{r}_n)$ be a homographic solution with zero angular momentum. We assume the centre of mass to be fixed at the origin. We can write $\mathbf{r}_i(t) = \lambda(t)A(t)\mathbf{a}_i$, $i = 1, \ldots, n$, with $\lambda(t) \in \mathbb{R}^+$ and $A(t) \in SO(3)$.

(a) Show that, with ω defined as in the proof of Theorem 7.1, we have

$$\sum_{i=1}^{n} m_i(\langle \mathbf{a}_i, \mathbf{a}_i \rangle \omega - \langle \omega, \mathbf{a}_i \rangle \mathbf{a}_i) = \mathbf{0}.$$

(b) By taking the scalar product of this equation with ω, deduce that $\omega = \mathbf{0}$, or the $\mathbf{a}_i$ are collinear and the line they determine coincides with the axis given by the instantaneous angular velocity ω. Conclude that the solution is homothetic.

7.7 Recall that a solution of the n-body problem where the configuration of the n bodies stays congruent to itself, as in the example shown in Figure 7.2, is called a relative equilibrium (Definition 6.34). We want to show that a relative equilibrium has to be planar, and the configuration rotates with constant angular velocity.

Let $(\mathbf{r}_1, \ldots, \mathbf{r}_n)$ be a relative equilibrium, with centre of mass fixed at the origin.

(a) If the motion is planar, we can write it as $\mathbf{r}_i(t) = \boldsymbol{\phi}(t)\mathbf{a}_i$ in the complex plane with $|\boldsymbol{\phi}| = 1$. Show that

$$\left|\sum_{i=1}^{n} m_i \mathbf{r}_i \times \dot{\mathbf{r}}_i\right| = |\dot{\boldsymbol{\phi}}| \cdot \sum_{i=1}^{n} m_i |\mathbf{a}_i|^2,$$

and conclude that $|\dot{\boldsymbol{\phi}}|$ is constant.

(b) In the general case, we can write $\mathbf{r}_i(t) = A(t)\mathbf{a}_i$ with $A(t) \in SO(3)$. Let Ω be the skew-symmetric matrix $A^{\mathrm{t}}\dot{A}$. Verify the identities

$$A^{\mathrm{t}}\ddot{A} = \dot{\Omega} + \Omega^2 \quad \text{and} \quad \Omega^2 = \frac{1}{2}(A^{\mathrm{t}}\ddot{A} + (A^{\mathrm{t}}\ddot{A})^{\mathrm{t}}).$$

(c) Show that the equations for the n-body problem take the form

$$A^{\mathrm{t}}\ddot{A}\mathbf{a}_i = \sum_{\substack{j=1 \\ j\neq i}}^{n} Gm_j \frac{\mathbf{a}_j - \mathbf{a}_i}{|\mathbf{a}_j - \mathbf{a}_i|^3}.$$

Conclude that, if the configuration $(\mathbf{a}_1, \dots, \mathbf{a}_n)$ is not planar, the matrix $A^{\mathrm{t}}\ddot{A}$ is constant.

Hint: If $\mathbf{a}_1, \mathbf{a}_2, \mathbf{a}_3$, say, are linearly independent, the 3×3 matrix with these vectors as columns is invertible.

(d) From (b) and (c) we have that Ω^2 is a constant matrix if the configuration $(\mathbf{a}_1, \dots, \mathbf{a}_n)$ is not planar (i.e. if the motion is not co-planar). By expressing Ω^2 in terms of the components $\omega_1, \omega_2, \omega_3$ of the instantaneous angular velocity ω, show that ω must likewise be constant.

(e) We may choose our cartesian coordinate system such that the instantaneous angular velocity takes the form $\omega = (0, 0, \gamma)^{\mathrm{t}}$ with γ a positive real number. Show that, if $A(0) = E$, then

$$A(t) = \begin{pmatrix} \cos(\gamma t) & -\sin(\gamma t) & 0 \\ \sin(\gamma t) & \cos(\gamma t) & 0 \\ 0 & 0 & 1 \end{pmatrix},$$

i.e. A is a rotation about the z-axis with constant angular velocity γ.

(f) Explain why the $\mathbf{a}_i$ must then in fact be contained in the xy-plane, so that the motion is planar after all.

Hint: Assuming that the $\mathbf{a}_i$ are not all contained in the xy-plane, consider the forces on the body whose position vector $\mathbf{r}_i$ has the largest z-component (in absolute value).

(g) This leaves the case where the motion is co-planar but not, *a priori*, planar. Here one can argue as in the proof of Theorem 7.1. Better, you may try to adapt the foregoing argument to this case.

7.8 Let $a_0, \dots, a_4$ be positive real numbers. Show that the polynomial

$$p(x) = x^5 + a_4x^4 + a_3x^3 - a_2x^2 - a_1x - a_0$$

has a unique zero on the set $\{x > 0\}$. This result, which we used to derive Euler's collinear solutions, is a special case of Descartes's sign rule, according to which the number of zeros on $\{x > 0\}$ of any given real polynomial is bounded from above by the number of sign changes in the coefficients.

Hint: Consider the derivatives p', p'', p'''.

7.9 Let $t \mapsto \mathbf{r}_i(t)$, $i = 1, 2, 3$, be a collinear solution of the three-body problem. It will be assumed that the line containing the three bodies at any given time does *not* stay fixed in space. The centre of mass will be assumed fixed at the origin. We now wish to derive the general form of these collinear solutions.

As in Exercise 7.4 we write the $\mathbf{r}_i$ in the form $\mathbf{r}_i(t) = A(t)(x_i(t), 0, 0)^{\mathrm{t}}$, where A has been chosen such that $\omega_A(t) = (0, q(t), r(t))^{\mathrm{t}}$.

(a) Show that

$$2r\dot{x}_i + \dot{r}x_i = 0 \text{ and } 2q\dot{x}_i + \dot{q}x_i = 0.$$

(b) Deduce the existence of two real constants α, β, not both equal to zero, such that

$$\alpha q + \beta r = 0.$$

Conclude further that there is a matrix $B \in \mathrm{SO}(3)$, independent of time, such that $\mathbf{r}_i(t) = A(t)B(x_i(t), 0, 0)^{\mathrm{t}}$ and $\omega_{AB}(t) = (0, 0, \tilde{r}(t))^{\mathrm{t}}$. This precomposition with B corresponds to a change of cartesian coordinates. In other words, we may assume without loss of generality that A has been chosen from the start such that $\omega_A(t) = (0, 0, r(t))^{\mathrm{t}}$.

Argue further that the solution is planar and the $\mathbf{r}_i$ can be written as

$$\mathbf{r}_i(t) = x_i(t) \begin{pmatrix} \cos\varphi(t) \\ \sin\varphi(t) \end{pmatrix}$$

with $\dot{\varphi} = r$.

(c) Using this explicit form of the $\mathbf{r}_i$, show that the size c of the angular momentum is given by $c = \lambda^2\dot{\varphi}$ with $\lambda(t) := \sqrt{\sum_{i=1}^{3} m_i x_i^2(t)}$. Since c is constant, this implies either $\varphi = 0$ – this is the case of a motion on a fixed line, which we have excluded – or $\dot{\varphi}(t) \neq 0$ for all t.

(d) Conclude from $c = \lambda^2\dot{\varphi}$ and the equations in (a) – observing that $\dot{\varphi} = r$ – that there are constants $a_i \in \mathbb{R}$ such that $x_i(t) = a_i\lambda(t)$. Show that λ satisfies the differential equation

$$\lambda^2\ddot{\lambda} - \frac{c^2}{\lambda} = -\mu$$

for some constant $\mu > 0$.

(e) Without loss of generality, assume that $a_1 < a_2 < a_3$. Show that the following relations hold between the constants a_i and μ:

$$\left.\begin{array}{rcl} -\mu a_1 & = & G\Big(\ \frac{m_2}{(a_2-a_1)^2}+\frac{m_3}{(a_3-a_1)^2}\Big) \\ -\mu a_2 & = & G\Big(-\frac{m_1}{(a_2-a_1)^2}+\frac{m_3}{(a_3-a_2)^2}\Big) \\ -\mu a_3 & = & G\Big(-\frac{m_1}{(a_3-a_1)^2}-\frac{m_2}{(a_3-a_2)^2}\Big). \end{array}\right\}$$

Verify that the choice of $a_2 - a_1$ determines the constants a_i and μ. Explain why this implies that any solution $t \mapsto \lambda(t)(\cos\varphi(t), \sin\varphi(t))$ of the Kepler problem determines a collinear solution of the three-body problem.

7.10 Describe Euler's collinear solution of the three-body problem explicitly for the case $m_1 = m_2 = m_3$.

7.11 (a) Show that with $\rho_i := |\mathbf{z} - \mathbf{z}_i|$, $i = 1, 2$, the potential function Φ in (PCR3B) can be written as

$$\Phi(\mathbf{z}) = (1-\mu)\left(\frac{\rho_1^2}{2} + \frac{1}{\rho_1}\right) + \mu\left(\frac{\rho_2^2}{2} + \frac{1}{\rho_2}\right).$$

This explains why the choice of additive constant in the original definition of Φ is apposite.

(b) Use (a) to show that Φ attains its absolute minimum in the Lagrange points L_4, L_5, and that the minimal value of Φ equals $3/2$.

(c) Sketch Hill's regions for the Jacobi constant taking values in the various intervals determined by the values of 2Φ at the libration points. In particular, explain why a motion from a neighbourhood of one of the primaries to a neighbourhood of the other is possible only for $C \leq 2\Phi(L_1)$.

7.12 The two Lagrange points correspond in an obvious manner to the homographic solutions of the three-body problem described in Section 7.1. Show that, as one should expect, the three collinear solutions with $m_3 = 0$ are the ones where the massless body sits at one of the three Euler points. Notice that, when we derived the Euler solutions, we were free to choose the order of the three bodies on the line, so for given primaries of mass m_1, m_2 there are three possible positions for m_3. In Section 7.2 it was merely for notational convenience that m_2 was assumed to sit between m_1 and m_3.

7.13 (a) Determine the coordinates of the five libration points in (PCR3B) for the case of the primaries having equal masses, i.e. $\mu = 1/2$.

(b) Compute the coordinates of the Lagrange points in the general case.

(c) Show that, for $0 < \mu < 1/2$, the Euler point L_1 lies closer to the smaller body $m_2 = \mu$ than to $m_1 = 1 - \mu$.

8

The differential geometry of the Kepler problem

> Then let them learn the use of the globes, which they will do in three weeks' time, and next let them fall upon spherical trigonometry; and here they will be ravished with celestial pleasure.
>
> John Aubrey, *An Idea of Education of Young Gentlemen*

In this chapter we return to the Kepler problem, which we are now going to study with a variety of geometric techniques. We shall encounter geometric transformations such as inversions and polar reciprocation, new types of spaces such as hyperbolic space or the projective plane, and differential geometric concepts such as geodesics and curvature.

In Section 8.1 we prove a theorem due to Hamilton, which says that the velocity vector of a Kepler solution traces out a circle (or an arc of a circle).

A remarkable consequence of this observation will be described in Sections 8.3 and 8.5 (for the elliptic case and the hyperbolic/parabolic case, respectively). By a geometric transformation known as inversion, one can set up a one-to-one correspondence between the velocity curves of Kepler solutions and geodesics, i.e. locally shortest curves, in spherical, hyperbolic or euclidean geometry. Not only does this permit a unified view of all solutions of the Kepler problem, including the regularised collision solutions, it also gives a natural interpretation of the eccentric anomaly as the arc length parameter of the corresponding geodesic. Sections 8.2 on inversion and 8.4 on hyperbolic geometry provide the necessary geometric background.

In Section 8.6, Hamilton's theorem is taken as the basis for an alternative proof of Kepler's first law. The geometric concepts behind this proof include polar reciprocals and the curvature of planar curves. In Section 8.6.5 I introduce the projective plane, which is the apposite setting for polar reciprocation.

In Section 8.7 it is shown that suitable holomorphic transformations of the complex plane lead to a duality between Kepler solutions on the one hand and solutions of Hooke's law for springs on the other. Since the latter are much easier to determine, this leads to yet another proof of Kepler's first law. A more geometric interpretation of this duality will follow in Section 9.1.

8.1 Hamilton's hodograph theorem

Definition 8.1 The **hodograph** of a C^1-curve $t \mapsto \mathbf{r}(t) \in \mathbb{R}^n$ is the image of its **velocity curve** $t \mapsto \mathbf{v}(t) := \dot{\mathbf{r}}(t)$.

In the cases we consider, the curve $t \mapsto \mathbf{v}(t)$ will traverse the hodograph in a single direction, so we shall think of the hodograph as an *oriented* curve.

We consider a solution $\mathbf{r}$ of the Kepler problem

$$\ddot{\mathbf{r}} = -\frac{\mu}{r^2} \cdot \frac{\mathbf{r}}{r} \tag{K}$$

with non-zero angular momentum $\mathbf{c} = \mathbf{r} \times \dot{\mathbf{r}}$. The main result of this section is the following theorem, see Figures 8.1 and 8.2.

Theorem 8.2 (Hamilton) *For solutions of the Kepler problem with non-vanishing angular momentum, the hodograph is an open subset in a circle in the plane of motion. Any oriented circle in $\mathbb{R}^3$ contained in a radial plane*[1] *can arise in this way. Conversely, from this oriented circle one can determine the hodograph and the corresponding solution of the Kepler problem (up to a time translation).*

We are going to make this statement more precise in a sequence of propositions, which together constitute a proof of the theorem.

Proposition 8.3 *Let $\mathbf{r}$ be a solution of* (K) *with $c \neq 0$. Then the velocity vector $\mathbf{v} = \dot{\mathbf{r}}$ moves on a circle C in the plane of motion. The radius of C equals μ/c, and the centre of C lies at distance $e\mu/c$ from the force centre $\mathbf{0}$, where e is the eccentricity of the conic section traced out by $\mathbf{r}$. Thus, the following statements hold.*

(e) *In the elliptic case, $\mathbf{0}$ lies in the interior of C.*
(p) *In the parabolic case, we have $\mathbf{0} \in C$.*
(h) *In the hyperbolic case, $\mathbf{0}$ is exterior to C.*

[1] A **radial plane** in $\mathbb{R}^3$ is a plane containing the origin.

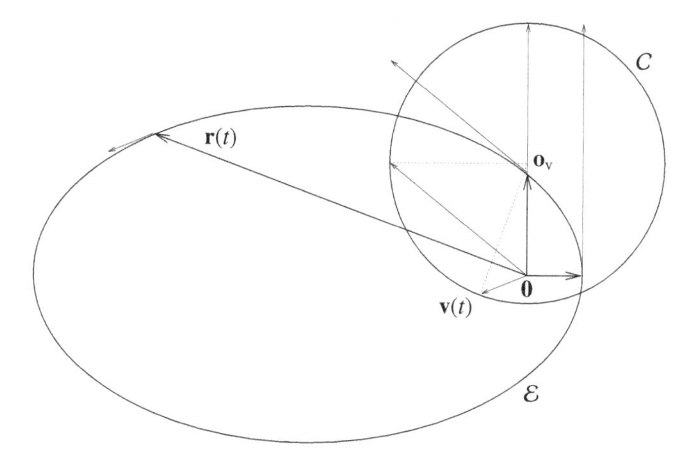

Figure 8.1 Hamilton's hodograph (elliptic case).

Proof Choose cartesian coordinates in $\mathbb{R}^3$ such that the angular momentum $\mathbf{c}$ is of the form $\mathbf{c} = (0, 0, c)$ with $c > 0$. This implies

$$\mathbf{r}(t) = r(t) \cdot (\cos\theta(t), \sin\theta(t), 0)$$

with $c = r^2\dot{\theta}$ (see the proof of Theorem 1.6). Hence $\dot{\theta} > 0$, which means that we may regard t as a function of θ.

Equation (K) can be written as

$$\dot{\mathbf{v}} = -\frac{\mu}{r^2}(\cos\theta, \sin\theta, 0).$$

Then

$$\frac{\mathrm{d}\mathbf{v}}{\mathrm{d}\theta} = \frac{\mathrm{d}\mathbf{v}}{\mathrm{d}t} \cdot \frac{\mathrm{d}t}{\mathrm{d}\theta} = \frac{\mathrm{d}\mathbf{v}}{\mathrm{d}t} \cdot \left(\frac{\mathrm{d}\theta}{\mathrm{d}t}\right)^{-1} = -\frac{\mu}{c}(\cos\theta, \sin\theta, 0).$$

By integration we obtain

$$\mathbf{v}(\theta) = \frac{\mu}{c}(-\sin\theta, \cos\theta, 0) + \mathbf{o}_\mathrm{v},$$

where $\mathbf{o}_\mathrm{v}$ is a constant of integration. This shows that $\mathbf{v}$ moves on a circle C about $\mathbf{o}_\mathrm{v}$ of radius μ/c.

Since $\mathbf{v}$ always lies in the orbital plane, we must have $\mathbf{o}_\mathrm{v} \in \mathbb{R}^2 \times \{0\} \subset \mathbb{R}^3$. We may assume that the cartesian coordinates in the orbital plane have been chosen such that $\mathbf{o}_\mathrm{v}$ lies on the positive y-axis. Write

$$\mathbf{o}_\mathrm{v} = (0, e\mu/c, 0)$$

with a suitable $e \geq 0$. Then

$$\mathbf{v}(\theta) = \frac{\mu}{c}(-\sin\theta, e + \cos\theta, 0). \tag{8.1}$$

It follows that

$$\begin{aligned} c = |\mathbf{r} \times \mathbf{v}| &= \frac{\mu r}{c}(e\cos\theta + \cos^2\theta + \sin^2\theta) \\ &= \frac{\mu r}{c}(1 + e\cos\theta), \end{aligned}$$

and hence

$$r = \frac{c^2/\mu}{1 + e\cos\theta}, \tag{8.2}$$

which we recognise as the description of a conic section in polar coordinates. Thus, we have once again proved Kepler's first law. Moreover, we see that the constant e, which we defined via the position of $\mathbf{o}_\mathrm{v}$, is actually the eccentricity of the conic section traced out by $\mathbf{r}$. □

With the choice of cartesian coordinates as in the preceding proof, the angle θ is the true anomaly, i.e. the eccentricity vector $\mathbf{e}$ points in the positive x-direction. Observe that $\mathbf{v}(\theta) - \mathbf{o}_\mathrm{v}$ is always orthogonal to $\mathbf{r}(\theta)$. This implies that the angle between two velocity vectors, seen from $\mathbf{o}_\mathrm{v}$, equals the angle between the corresponding position vectors. See the dotted lines in Figure 8.1 and the angle marked θ in Figure 8.2.

I shall refer to the oriented circle C determined by the hodograph of $\mathbf{r}$ as the **velocity circle** of $\mathbf{r}$.

Lemma 8.4 *Any oriented circle in $\mathbb{R}^3$ contained in a radial plane arises as the velocity circle of a suitable solution of the Kepler problem.*

Proof It is implicit in the proof of Proposition 8.3 that the velocity circle of the solution (8.2) of the Kepler problem (contained in the xy-plane and with eccentricity vector pointing in the positive x-direction) is a circle of radius μ/c, centred at $(0, e\mu/c, 0)$, and traversed counter-clockwise. In this way, any positively oriented circle (in the xy-plane) centred on the positive y-axis can occur. If $\mathbf{r}$ is a solution of the Kepler problem, then so is $A\mathbf{r}$ for any $A \in \mathrm{O}(3)$, see Exercise 1.1. If C is the velocity circle of $\mathbf{r}$, then AC is the velocity circle of $A\mathbf{r}$. In this way, any oriented circle in $\mathbb{R}^3$ contained in a radial plane can be realised. □

In the following proposition we write h for the energy of $\mathbf{r}$ per unit mass, as in Chapter 3.

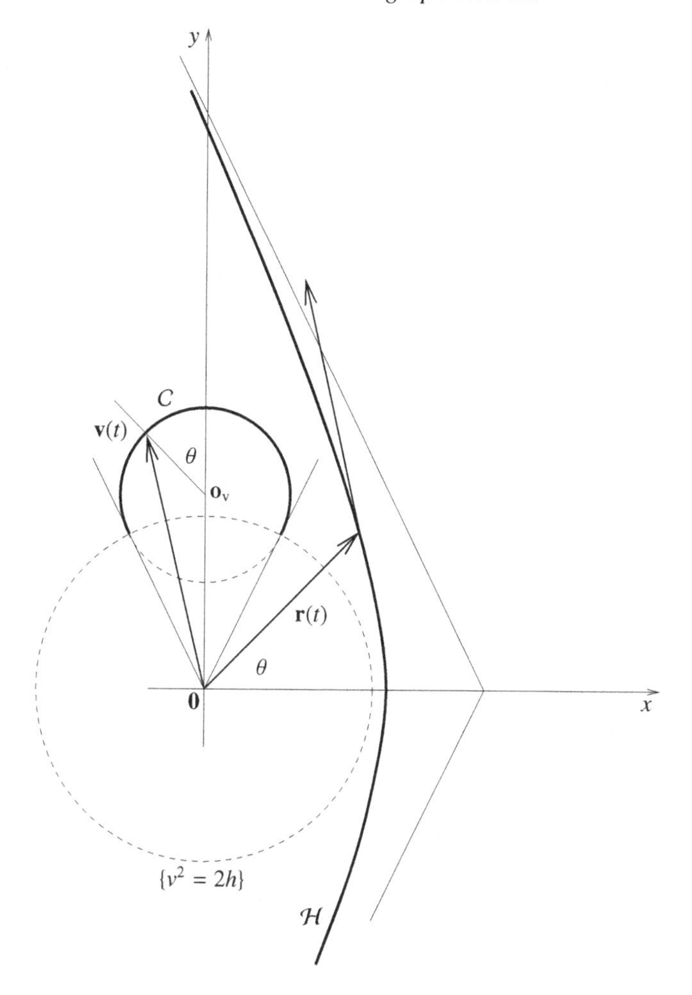

Figure 8.2 Hamilton's hodograph (hyperbolic case).

Proposition 8.5 *Let C be the velocity circle of the solution $\mathbf{r}$ of* (K). *The velocity curve $t \mapsto \mathbf{v}(t)$ traverses exactly that part of C where $v^2 > 2h$. This means that, in the elliptic case (where $h < 0$), the full velocity circle is traversed; in the parabolic case (where $h = 0$), the complement of $\mathbf{0} \in C$ is traversed.*

Proof As shown in Corollary 3.3, in the case $h < 0$ (i.e. $e < 1$) the ellipse is traversed infinitely often by the curve $t \mapsto \mathbf{r}(t)$. It follows that the velocity curve $t \mapsto \mathbf{v}(t)$ likewise traverses the velocity circle infinitely often.

For $h \geq 0$ (i.e. $e \geq 1$), we showed that the angle $\theta(t)$ traverses the full interval determined by the condition $1 + e\cos\theta > 0$. For $e = 1$ this means the interval $(-\pi, \pi)$, which corresponds to $C \setminus \{\mathbf{0}\} = C \cap \{v^2 > 0\}$, see Figure 8.3.

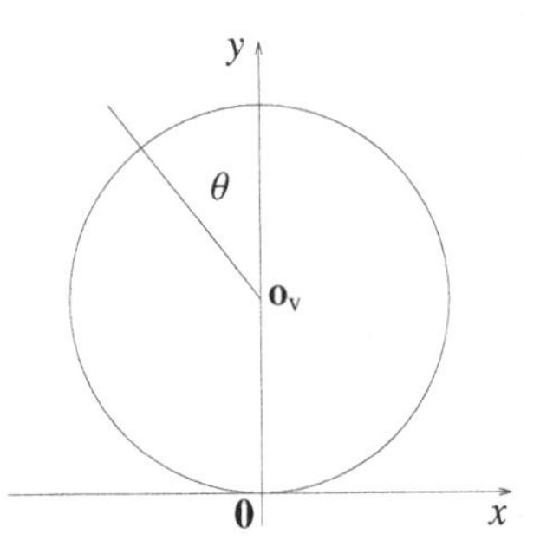

Figure 8.3 The velocity circle in the parabolic case.

For $e > 1$ we argue with the energy formula $h = v^2/2 - \mu/r$, see (3.6). This implies $v^2 > 2h$. From Corollary 3.3 we know that $r \to \infty$ for $t \to \pm\infty$, so the velocity curve does indeed traverse exactly that part of C where this inequality holds. □

We now want to show how to recover the hodograph and the solution of the Kepler problem from its velocity circle. Given an oriented circle C contained in a radial plane in $\mathbb{R}^3$, we can rotate our cartesian coordinate system in such a way that C becomes a positively oriented circle in the xy-plane with centre on the positive y-axis. This choice of coordinates corresponds to a solution of (K) with eccentricity vector pointing in the positive x-direction. Write r_C for the radius of C, and $\mathbf{o}_C = (0, |\mathbf{o}_C|, 0)$ for its centre. By the proof of Proposition 8.3, we find the angular momentum $\mathbf{c} = (0, 0, c)$ and the eccentricity e of the corresponding Kepler solution $\mathbf{r}$ from this information via

$$c = \mu/r_C \quad \text{and} \quad e = |\mathbf{o}_C|/r_C.$$

These quantities determine $\mathbf{r}$ up to a time translation, as our results in Chapters 3 and 4 have shown.

In order to complete the proof of Theorem 8.2, it remains to show that the velocity circle C determines the hodograph. Of course this is only an issue in the hyperbolic case (i.e. in the case where the radius of the velocity circle is smaller than the distance of its centre from the origin). Analytically, one can compute h, and hence the hodograph $C \cap \{v^2 > 2h\}$, from c and e via equation (3.7). But there is a nicer geometric way to determine the hodograph, which will be relevant later on.

Figure 8.2 seems to suggest that the 'energy circle' $\{v^2 = 2h\}$ intersects the velocity circle C orthogonally. This is indeed implied by the following lemma. So the hodograph can be found by drawing the two tangents from $\mathbf{0}$ to C.

Lemma 8.6 *Let C be a circle in $\mathbb{R}^2$ with centre $\mathbf{o}_C$ and radius r_C. Let ℓ be a radial line in $\mathbb{R}^2$ intersecting C in one or two points. Write $\mathbf{v}_1, \mathbf{v}_2$ for the intersection points, with $\mathbf{v}_1 = \mathbf{v}_2$ in the case of a single point of tangential intersection. Then*

$$\langle \mathbf{v}_1, \mathbf{v}_2 \rangle = |\mathbf{o}_C|^2 - r_C^2.$$

If C is the velocity circle of a solution of the Kepler problem, this quantity equals $2h$.

Proof Let $\mathbf{w}$ be the point on C diametrically opposite to $\mathbf{v}_1$. Figure 8.4 shows the various relative positions of C, ℓ and the origin $\mathbf{0}$. Write $\mathbf{v}_1 = \mathbf{o}_C + \mathbf{s}$ and $\mathbf{w} = \mathbf{o}_C - \mathbf{s}$ with $|\mathbf{s}| = r_C$, and $\mathbf{v}_2 = \mathbf{o}_C + \mathbf{t}$ with $|\mathbf{t}| = r_C$. The identity $\langle \mathbf{s} - \mathbf{t}, \mathbf{s} + \mathbf{t} \rangle = 0$ translates into $\langle \mathbf{v}_1 - \mathbf{v}_2, \mathbf{v}_2 - \mathbf{w} \rangle = 0$; this is the theorem of Thales. Since $\mathbf{v}_1$ and $\mathbf{v}_2$ lie on a radial line ℓ, for $\mathbf{v}_1 \neq \mathbf{v}_2$ we know that $\mathbf{v}_1$ is a multiple of $\mathbf{v}_1 - \mathbf{v}_2$, and it follows that

$$\langle \mathbf{v}_1, \mathbf{v}_2 - \mathbf{w} \rangle = 0. \tag{8.3}$$

If $\mathbf{v}_1 = \mathbf{v}_2$, i.e. in the case of a tangential intersection, we have $\mathbf{v}_1 \perp \mathbf{s}$ and $\mathbf{v}_2 - \mathbf{w} = \mathbf{s} + \mathbf{t} = 2\mathbf{s}$. Once again, (8.3) holds. Hence

$$\langle \mathbf{v}_1, \mathbf{v}_2 \rangle = \langle \mathbf{v}_1, \mathbf{w} \rangle = \langle \mathbf{o}_C + \mathbf{s}, \mathbf{o}_C - \mathbf{s} \rangle = |\mathbf{o}_C|^2 - r_C^2.$$

In the case of a velocity circle, we have $|\mathbf{o}_C| = |\mathbf{o}_\mathrm{v}| = e\mu/c$ and $r_C = \mu/c$, hence

$$\langle \mathbf{v}_1, \mathbf{v}_2 \rangle = \frac{\mu^2}{c^2}(e^2 - 1) = 2h$$

by (3.7). □

In Exercise 8.27 you will be asked to show that the circularity of the hodograph characterises the Newtonian law of attraction.

8.2 Inversion and stereographic projection

In this section we study some geometric transformations of euclidean space $\mathbb{R}^n$. In the next section we shall see how these transformations relate to Hamilton's hodograph.

By $S^{n-1} \subset \mathbb{R}^n$ I always denote the unit sphere in $\mathbb{R}^n$, i.e. the sphere of radius 1 centred at the origin. In this section, however, we want to consider spheres of arbitrary centre and radius. I shall use the notation $\mathbb{S}$ for such spheres.

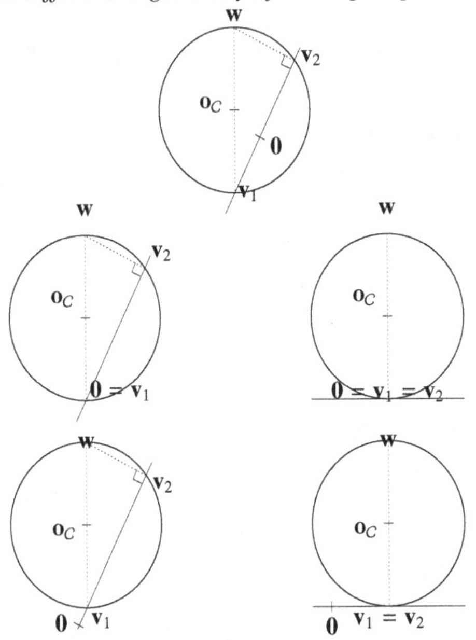

Figure 8.4 Proof of Lemma 8.6.

For the definition of our first geometric transformation we need to choose a reference sphere $\mathbb{S}_0$. Fix a point $\mathbf{p}_0 \in \mathbb{R}^n$ and a radius $r_0 > 0$, and let

$$\mathbb{S}_0 := \{\mathbf{x} \in \mathbb{R}^n : |\mathbf{x} - \mathbf{p}_0| = r_0\}$$

be the sphere of radius r_0 about $\mathbf{p}_0$.

Definition 8.7 The **inversion in** $\mathbb{S}_0$ is the smooth map

$$\Phi \colon \mathbb{R}^n \setminus \{\mathbf{p}_0\} \longrightarrow \mathbb{R}^n \setminus \{\mathbf{p}_0\}$$

determined by the following requirements, see Figure 8.5[2]:

(i) The point $\Phi(\mathbf{x})$ lies on the ray from $\mathbf{p}_0$ through $\mathbf{x}$.
(ii) $|\Phi(\mathbf{x}) - \mathbf{p}_0| \cdot |\mathbf{x} - \mathbf{p}_0| = r_0^2$.

Explicitly, we have

$$\Phi(\mathbf{x}) = \mathbf{p}_0 + r_0^2 \frac{\mathbf{x} - \mathbf{p}_0}{|\mathbf{x} - \mathbf{p}_0|^2}.$$

We notice the following properties of the inversion Φ.

- The inversion Φ is an **involution** of $\mathbb{R}^n \setminus \{\mathbf{p}_0\}$, that is, $\Phi \circ \Phi$ is the identity.

[2] See Figure 8.13 for a geometric construction of the inverse point.

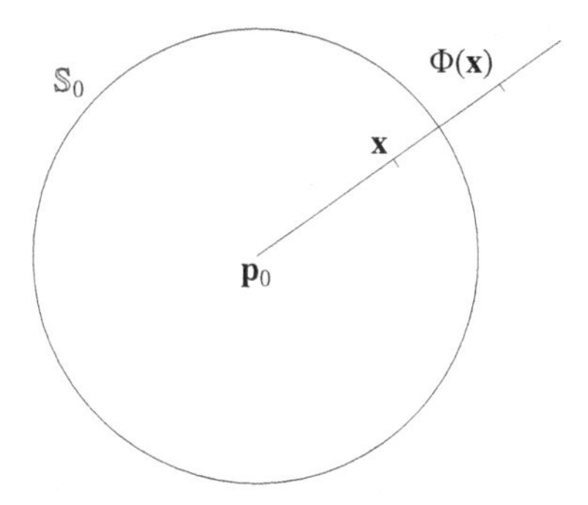

Figure 8.5 Inversion in $\mathbb{S}_0$.

- The fixed points of Φ are precisely the points on $\mathbb{S}_0$.

Next we define stereographic projection for arbitrary spheres and hyperplanes not passing through a chosen centre of projection.

Definition 8.8 Let $\mathbb{S} \subset \mathbb{R}^n$ be an $(n-1)$-sphere, $\mathbf{p}, \mathbf{q} \in \mathbb{S}$ a pair of antipodal points and $E \subset \mathbb{R}^n$ a hyperplane orthogonal to $\mathbf{p} - \mathbf{q}$ not passing through $\mathbf{p}$. The **stereographic projection** of $\mathbb{S}$ from $\mathbf{p}$ onto E is the map

$$\Psi\colon\ \mathbb{S} \setminus \{\mathbf{p}\} \longrightarrow E,$$

where $\Psi(\mathbf{x})$ is the unique point in E on the line through $\mathbf{p}$ and $\mathbf{x}$, see Figure 8.6.

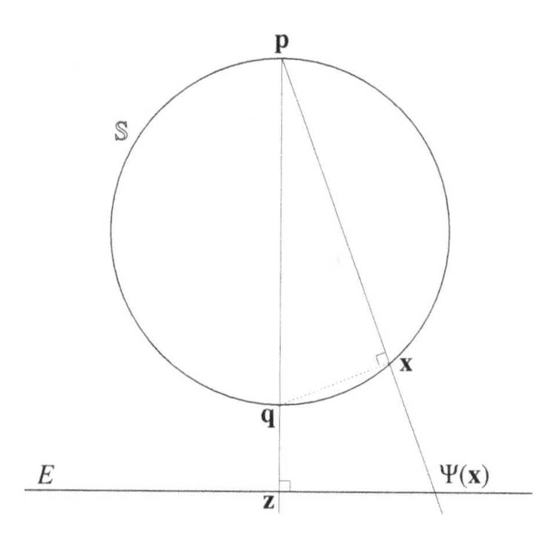

Figure 8.6 Stereographic projection of $\mathbb{S}$ from $\mathbf{p}$ onto E.

Lemma 8.9 *Stereographic projection is, up to sign, inversion of $\mathbb{S}$ in a suitable sphere $\mathbb{S}_0$.*

Proof Let $\mathbf{z} \neq \mathbf{p}$ be the intersection point of E with the line through $\mathbf{p}$ and $\mathbf{q}$, see Figure 8.6. By the theorem of Thales, the triangles $\triangle\mathbf{pqx}$ and $\triangle\mathbf{p}\Psi(\mathbf{x})\mathbf{z}$ are similar right-angled triangles. It follows that

$$|\mathbf{x} - \mathbf{p}| \cdot |\Psi(\mathbf{x}) - \mathbf{p}| = |\mathbf{p} - \mathbf{z}| \cdot |\mathbf{p} - \mathbf{q}| =: r^2.$$

If $\mathbf{z}$ lies on the same side of $\mathbf{p}$ as $\mathbf{q}$ along the line $\mathbf{pq}$, then Ψ is the inversion in the sphere of radius r about $\mathbf{p}$; if $\mathbf{z}$ lies on the other side, Ψ is minus the inversion. □

For the formulation of the following proposition it is convenient to add a single point ∞ 'at infinity' to $\mathbb{R}^n$. Inversion in $\mathbb{S}_0 \subset \mathbb{R}^n$ can then be extended to a bijective map[3]

$$\Phi\colon\ \mathbb{R}^n \cup \{\infty\} \longrightarrow \mathbb{R}^n \cup \{\infty\}$$

by setting $\Phi(\mathbf{p}_0) := \infty$ and $\Phi(\infty) := \mathbf{p}_0$. Furthermore, by a **$k$-dimensional plane** in $\mathbb{R}^n \cup \{\infty\}$ we shall mean a k-dimensional affine subspace of $\mathbb{R}^n$ with the point ∞ added. For a discussion of **k-dimensional spheres** in $\mathbb{R}^n$ see Exercise 8.2.

A C^1-map between open subsets of $\mathbb{R}^n$ is called **conformal** if it preserves angles. This means that if two C^1-curves in the domain of definition intersect at some point with non-zero velocity, then the angle between their tangent lines is the same as the angle between the corresponding tangent lines of the image curves.

Proposition 8.10 (Steiner) *The inversion Φ of $\mathbb{R}^n \cup \{\infty\}$ in $\mathbb{S}_0$ sends k-dimensional spheres and planes to spheres or planes. More precisely, the images of spheres and planes are as follows:*

$$\begin{aligned} \textit{plane through } \mathbf{p}_0 &\longrightarrow \textit{plane through } \mathbf{p}_0 \\ \textit{sphere through } \mathbf{p}_0 &\longleftrightarrow \textit{plane not containing } \mathbf{p}_0 \\ \textit{sphere not containing } \mathbf{p}_0 &\longrightarrow \textit{sphere not containing } \mathbf{p}_0 \end{aligned}$$

The restriction of Φ to $\mathbb{R}^n \setminus \{\mathbf{p}_0\}$ is a conformal transformation.

With Lemma 8.9 the following is immediately apparent, see also Exercise 8.5.

[3] In Exercise 8.3 it will be shown how to put a topology on the space $\mathbb{R}^n \cup \{\infty\}$. One can then discuss the continuity of the extended map Φ; this too will be done in Exercise 8.3, for the analogous case of the stereographic projection.

Corollary 8.11 *Stereographic projection $\Psi\colon\ \mathbb{S}\setminus\{\mathbf{p}\} \to E$ is a conformal map that sends any circle in $\mathbb{S}$ (that is, an intersection of $\mathbb{S}$ with an affine 2-plane) to a circle or a line in E.* □

Proof of Proposition 8.10 For a plane E through $\mathbf{p}_0$, the restriction of Φ to E is the inversion in $\mathbb{S}_0 \cap E$, which sends E to E.

For spheres through $\mathbf{p}_0$ and planes not containing $\mathbf{p}_0$ the statement follows by reversing the argument in the proof of Lemma 8.9, see Exercise 8.6.

For an $(n-1)$-sphere $\mathbb{S}$ not containing $\mathbf{p}_0$ we argue as in Lemma 8.6. Choose cartesian coordinates such that $\mathbf{p}_0 = \mathbf{0}$. Let $\mathbf{x}$ be a point in $\mathbb{S}$, and let $\mathbf{x}'$ be the second point of intersection of the radial line through $\mathbf{x}$ with $\mathbb{S}$, where we set $\mathbf{x}' = \mathbf{x}$ if this radial line intersects $\mathbb{S}$ tangentially at $\mathbf{x}$. By Lemma 8.6 we know that $|\mathbf{x}| \cdot |\mathbf{x}'| = \pm\langle \mathbf{x}, \mathbf{x}' \rangle$ is a constant (different from 0) as $\mathbf{x}$ varies over $\mathbb{S}$. Write this constant as r_0^2/a with $a \in \mathbb{R}^+$. Then $|\mathbf{x}| \cdot |a\mathbf{x}'| = r_0^2$, which implies that $\Phi(\mathbf{x}) = \pm a\mathbf{x}'$, with plus or minus sign depending on $\mathbf{0}$ being exterior or interior to $\mathbb{S}$. As $\mathbf{x}$ varies over all of $\mathbb{S}$, so does $\mathbf{x}'$; hence $\Phi(\mathbb{S}) = \pm a\mathbb{S}$, which is again an $(n-1)$-sphere.

Given a k-sphere not containing $\mathbf{p}_0$, we can think of it as the intersection of an $(n-1)$-sphere with a $(k+1)$-dimensional affine plane, at least one of which does not contain $\mathbf{p}_0$. By the cases already discussed, the image under Φ will then be the sphere of intersection between two spheres or a sphere with a plane.

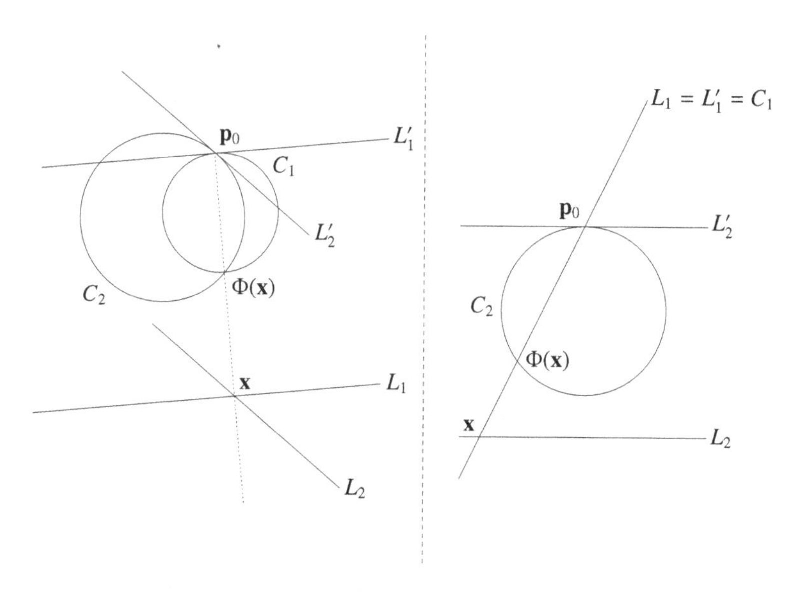

Figure 8.7 Conformality of inversions.

It remains to prove conformality of Φ. Consider two curves γ_1, γ_2 in $\mathbb{R}^n$ intersecting each other at some point $\mathbf{x} \neq \mathbf{p}_0$ with well-defined tangent lines $L_1 \neq L_2$ at that point. Let L_i' be the line through $\mathbf{p}_0$ parallel to L_i, and set $C_i = \Phi(L_i)$. Observe that, if $\mathbf{p}_0 \in L_i$, then $L_i = L_i' = C_i$; if $\mathbf{p}_0 \notin L_i$, then C_i is a circle in the plane determined by L_i and L_i'. Figure 8.7 shows a schematic picture of these two cases. If E_{12} is the plane determined by L_1 and L_2, then C_1 and C_2 lie in $\Phi(E_{12})$, which is the plane E_{12} for $\mathbf{p}_0 \in E_{12}$, and a 2-sphere otherwise.

By considering the inversion restricted to the plane defined by L_i and L_i' (for $\mathbf{p}_0 \notin L_i$), we see that C_i is tangent to L_i' at $\mathbf{p}_0$. Moreover, C_1 and C_2 intersect at the two points $\mathbf{p}_0$ and $\Phi(\mathbf{x})$ with equal angle, which is exactly the angle between the image curves $\Phi(\gamma_1), \Phi(\gamma_2)$ at $\Phi(\mathbf{x})$ that we wish to determine. Hence

$$\measuredangle_{\Phi(\mathbf{x})}(C_1, C_2) = \measuredangle_{\mathbf{p}_0}(C_1, C_2) = \measuredangle_{\mathbf{p}_0}(L_1', L_2') = \measuredangle_{\mathbf{x}}(L_1, L_2),$$

as we wanted to show. □

8.3 Spherical geometry and Moser's theorem

In order to apply the theory from the preceding section to the Kepler problem, we first need to derive an explicit description of the stereographic projection Ψ of the unit 3-sphere $S^3 \subset \mathbb{R}^4$ from the 'north pole' $N := (0, 0, 0, 1)$ onto the 'equatorial plane' $\mathbb{R}^3 \times \{0\}$; see Figure 8.8, which is lacking one dimension.

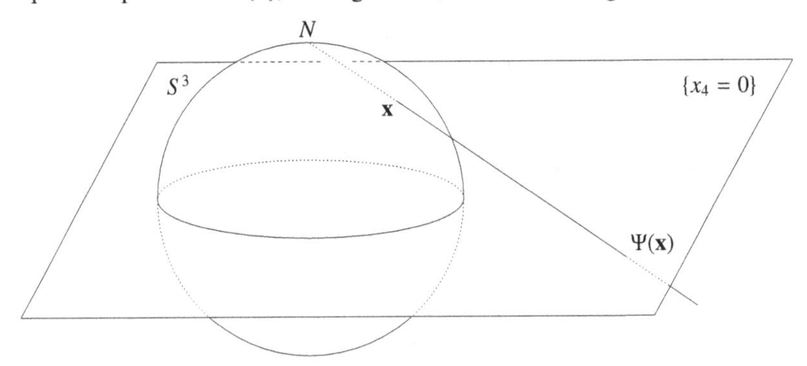

Figure 8.8 Stereographic projection $\Psi\colon\ S^3 \setminus \{N\} \to \{x_4 = 0\}$.

We write points on $S^3 \subset \mathbb{R}^4$ as $\mathbf{x} = (x_1, x_2, x_3, x_4)$, and points in $\mathbb{R}^3 \times \{0\} \equiv \mathbb{R}^3$ as $\mathbf{w} = (w_1, w_2, w_3)$. The 'south pole' on S^3 is $S := (0, 0, 0, -1)$.

The line through N and a point $\mathbf{x} \in S^3 \setminus \{N\}$ is given by

$$(1-s)\cdot(0,0,0,1) + s\cdot(x_1,x_2,x_3,x_4), \quad s \in \mathbb{R}.$$

So the point $\Psi(\mathbf{x}) \in \mathbb{R}^3$ is determined by the condition $(1-s)+sx_4 = 0$, or $s = 1/(1-x_4)$. It follows that Ψ is given by

$$\Psi(x_1,x_2,x_3,x_4) = \frac{1}{1-x_4}(x_1,x_2,x_3).$$

Lemma 8.12 *Under the diffeomorphism*[4] *between $S^3 \setminus \{N,S\}$ and $\mathbb{R}^3 \setminus \{\mathbf{0}\}$ given by the stereographic projection Ψ, the antipodal map $\mathbf{x} \mapsto -\mathbf{x}$ on $S^3 \setminus \{N,S\}$ corresponds to the negative inversion $\mathbf{w} \mapsto -\mathbf{w}/w^2$ in the unit sphere S^2, i.e. the following diagram of maps commutes:*

$$\begin{array}{ccc} S^3\setminus\{N,S\} & \xrightarrow{\mathbf{x}\mapsto-\mathbf{x}} & S^3\setminus\{N,S\} \\ {\scriptstyle\Psi}\downarrow & & \downarrow{\scriptstyle\Psi} \\ \mathbb{R}^3\setminus\{\mathbf{0}\} & \xrightarrow[\mathbf{w}\mapsto-\mathbf{w}/w^2]{} & \mathbb{R}^3\setminus\{\mathbf{0}\}. \end{array}$$

Proof Set $\mathbf{w} = \Psi(\mathbf{x})$. We compute

$$w^2 = \frac{x_1^2+x_2^2+x_3^2}{(1-x_4)^2} = \frac{1-x_4^2}{(1-x_4)^2} = \frac{1+x_4}{1-x_4}.$$

Hence

$$\Psi(-\mathbf{x}) = \frac{1}{1+x_4}(-x_1,-x_2,-x_3) = -\frac{1-x_4}{1+x_4}\mathbf{w} = -\mathbf{w}/w^2,$$

as we had to show. □

Remark 8.13 If we extend the negative inversion of $\mathbb{R}^3 \setminus \{\mathbf{0}\}$ to a homeomorphism of $\mathbb{R}^3 \cup \{\infty\}$ by sending $\mathbf{0}$ to ∞ and *vice versa*, and the stereographic projection $\Psi\colon S^3 \setminus \{N\} \to \mathbb{R}^3$ to a homeomorphism $S^3 \to \mathbb{R}^3 \cup \{\infty\}$ by sending N to ∞, the commutative diagram in Lemma 8.12 extends to

$$\begin{array}{ccc} S^3 & \xrightarrow{\mathbf{x}\mapsto-\mathbf{x}} & S^3 \\ {\scriptstyle\Psi}\downarrow & & \downarrow{\scriptstyle\Psi} \\ \mathbb{R}^3\cup\{\infty\} & \xrightarrow[\mathbf{w}\mapsto-\mathbf{w}/w^2]{} & \mathbb{R}^3\cup\{\infty\}. \end{array}$$

We now consider solutions of the Kepler problem with energy $h = -1/2$, including linear solutions, i.e. those with vanishing angular momentum. By Remark 3.5 we then have $a = \mu$; angular momentum and eccentricity are related

[4] For the differentiability of Ψ^{-1} see Exercise 8.7.

by $c = \mu\sqrt{1-e^2}$, see (3.7). It follows that elliptic solutions are determined up to a time translation by the orbital plane (with orientation given by the angular momentum $\mathbf{c}$) and the eccentricity vector $\mathbf{e}$ in this plane; linear solutions are determined by the eccentricity vector $\mathbf{e}$ (with $e = 1$). Recall from the proof of Theorem 3.1 that in the linear case we have $\mathbf{e} = -\mathbf{r}/r$. When the collision in such a linear solution is regularised as in Exercise 4.7, the hodograph is the full radial line containing the orbit, oriented by the eccentricity vector, including the point ∞ at the time of collision. Indeed, the linear solution with $a = \mu$ and $\mathbf{e} = (1, 0, 0)$ is given by

$$x(u) = -a(1 - \cos u), \quad t(u) = a(u - \sin u),$$

hence

$$\dot{x} = \frac{\mathrm{d}x}{\mathrm{d}u} \cdot \frac{\mathrm{d}u}{\mathrm{d}t} = \frac{\mathrm{d}x}{\mathrm{d}u} \cdot \left(\frac{\mathrm{d}t}{\mathrm{d}u}\right)^{-1} = \frac{-\sin u}{1 - \cos u}. \tag{8.4}$$

Beware that $u \mapsto x(u)$ and $u \mapsto t(u)$ are smooth functions on all of $\mathbb{R}$, but $t \mapsto x(t)$ and $t \mapsto u(t)$ are not differentiable at $t \in 2\pi\mathbb{Z}$. However, if we include the point ∞ in the space of possible velocities, $\dot{x}$ may be regarded as a continuous function of u on all of $\mathbb{R}$.

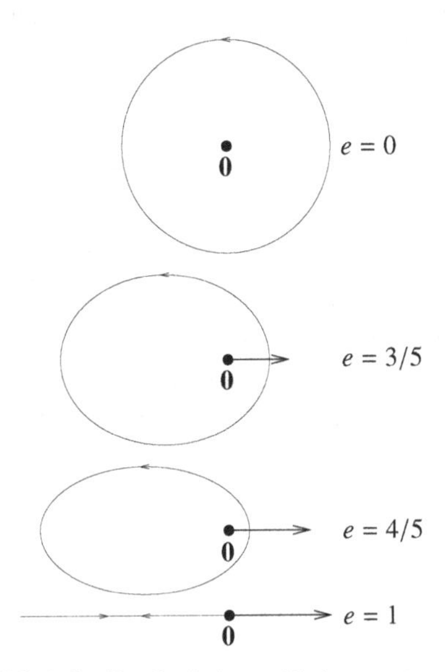

Figure 8.9 A family of solutions with the same a.

Figure 8.9 shows a family of solutions with the same a, starting at the circular solution ($e = 0$) and ending at a linear solution ($e = 1$). This family varies continuously with the parameter e. We see that within the same orbital plane the circular solution can be continuously deformed into elliptic solutions with eccentricity vector pointing in any direction. The linear solution can also be deformed into elliptic solutions oriented clockwise; this corresponds to a deformation of $\mathbf{c}$ via vectors staying on a fixed line, with the deformation passing through $\mathbf{c} = \mathbf{0}$ for the linear solution.

We now wish to make this continuous dependence of the Kepler solutions on $\mathbf{e}$ and $\mathbf{c}$ (subject to the relation $c = \mu\sqrt{1-e^2}$) geometrically more transparent. This can be achieved by relating the hodographs of Kepler solutions with $h = -1/2$ via stereographic projection to certain curves on S^3. The curves in question are the following.

Definition 8.14 A **great circle** on S^3 is the intersection of S^3 with a radial 2-plane in $\mathbb{R}^4$.

Remark 8.15 These great circles are the locally distance minimising curves on S^3, or so-called **geodesics** (when parametrised proportionally to arc length, i.e. with constant speed). We shall discuss geodesics in more detail in Section 8.4 in the context of hyperbolic geometry. For more on spherical geodesics see Exercises 8.8 to 8.10.

Theorem 8.16 (Moser) *The stereographic projection $\Psi\colon S^3 \to \mathbb{R}^3 \cup \{\infty\}$ yields a bijection between oriented great circles on S^3 and the hodographs of Kepler solutions in $\mathbb{R}^3$ with energy $h = -1/2$.*

Proof The great circles are precisely those circles on S^3 that are invariant under the antipodal map $\mathbf{x} \mapsto -\mathbf{x}$. Hence, by Corollary 8.11, Lemma 8.12 and Remark 8.13, Ψ maps the great circles to precisely those circles or lines in $\mathbb{R}^3 \cup \{\infty\}$ that are invariant under the negative inversion $\mathbf{w} \mapsto -\mathbf{w}/w^2$.

Any great circle in S^3 lies on a 2-sphere through the north pole. It follows that the invariant circles in $\mathbb{R}^3$ lie in radial planes. The invariant lines are simply the radial lines.

By the discussion above, each oriented radial line corresponds to a unique collision solution of the Kepler problem with $h = -1/2$.

It remains to show that the hodographs of Kepler solutions with $h = -1/2$ and $c \neq 0$ are precisely the invariant circles in $\mathbb{R}^3$. By Section 8.1, the hodographs are the circles in $\mathbb{R}^3$ contained in a radial plane and, by Lemma 8.6, such that $\langle \mathbf{v}_1, \mathbf{v}_2 \rangle = -1$ for any pair $\mathbf{v}_1, \mathbf{v}_2$ of intersection points with a radial line. The latter condition is equivalent to $\mathbf{v}_2 = -\mathbf{v}_1/v_1^2$, which is precisely the invariance condition. □

We now give a second, more computational proof of Theorem 8.16, which yields additional information about the respective parametrisations of great circles and hodographs.

To start with, we consider elliptic Kepler solutions with $\mathbf{c} \neq \mathbf{0}$ pointing in the positive z-direction (so the motion is counter-clockwise in the xy-plane) and $\mathbf{e} \neq \mathbf{0}$ (with $e < 1$) pointing in the positive x-direction, or $e = 0$. Under these assumptions, the true anomaly equals the angle θ between the position vector and the positive x-axis. By comparing the parametrisation of the orbit in terms of θ with that in terms of the eccentric anomaly u –

$$r(\cos\theta, \sin\theta) = a(\cos u - e, \sqrt{1-e^2}\sin u)$$

– we see that $r = a(1 - e\cos u)$ and

$$1 + e\cos\theta = \frac{1-e^2}{1-e\cos u},$$

see Exercises 4.1 and 4.2. Hence

$$\frac{\sin u}{1-e\cos u} = \frac{\sin\theta}{\sqrt{1-e^2}} \tag{8.5}$$

and

$$\begin{aligned}\frac{\sqrt{1-e^2}\cos u}{1-e\cos u} &= \frac{\sqrt{1-e^2}}{e}\left(\frac{1}{1-e\cos u} - 1\right)\\ &= \frac{\sqrt{1-e^2}}{e}\left(\frac{1+e\cos\theta}{1-e^2} - 1\right)\\ &= \frac{e+\cos\theta}{\sqrt{1-e^2}}.\end{aligned} \tag{8.6}$$

For this last computation we have assumed $e \neq 0$, but (8.6) holds trivially for $e = 0$.

Now consider the great circle of S^3 given by

$$\gamma(u) = (-\sin u, \sqrt{1-e^2}\cos u, 0, e\cos u),$$

where we allow e to take any value in the interval $[0, 1]$. Notice that this parametrisation of the great circle has unit speed.[5]

The image of γ under stereographic projection is

$$\Psi \circ \gamma(u) = \frac{1}{1-e\cos u}(-\sin u, \sqrt{1-e^2}\cos u, 0).$$

For $e = 1$ this gives

$$u \longmapsto \left(\frac{-\sin u}{1-\cos u}, 0, 0\right),$$

[5] We also speak of **arc length parametrisation**.

which by (8.4) we recognise as the velocity curve of the linear solution with $h = -1/2$ and $\mathbf{e} = (1, 0, 0)$.

For $0 \leq e < 1$ we substitute θ for u from the identities (8.5) and (8.6), to obtain the curve

$$\theta \longmapsto \frac{1}{\sqrt{1 - e^2}}(-\sin\theta, e + \cos\theta, 0).$$

By comparing this with the parametrisation of the velocity circle found in (8.1), we see that this is the hodograph of the Kepler solution for $h = -1/2$, $\mathbf{e} = (e, 0, 0)$ and $\mathbf{c} = (0, 0, \mu\sqrt{1 - e^2})$.

I now want to summarise these considerations in a single comprehensive statement about the relation between great circles on S^3 and velocity curves of Kepler solutions in $\mathbb{R}^3$. Under stereographic projection, radial planes in $\mathbb{R}^3$ correspond to 2-spheres through N in S^3. If we fix the orbital plane to be the xy-plane (and hence the velocity curve to lie in the w_1w_2-plane), the corresponding 2-sphere is $S^3 \cap \{x_3 = 0\}$.

Define $\boldsymbol{\gamma}_\alpha$, $\alpha \in \mathbb{R}$, to be the family of great circles of S^3 with arc length parametrisation given by

$$\boldsymbol{\gamma}_\alpha(u) = (-\sin u, \cos\alpha\cos u, 0, \sin\alpha\cos u).$$

These are the great circles in $S^2 = S^3 \cap \{x_3 = 0\}$ obtained by rotating the unit circle in the x_1x_2-plane about the x_1-axis.

Above we considered those Kepler solutions with $h = -1/2$ that have $\mathbf{e} = (e, 0, 0)$ and $\mathbf{c} = (0, 0, \mu\sqrt{1 - e^2})$, $0 \leq e \leq 1$. As we have seen, the velocity curve of such a solution corresponds under stereographic projection to the great circle $\boldsymbol{\gamma}_\alpha$ with $\sin\alpha = e$ and $\mu\cos\alpha = c$.

Similar considerations apply when we change the sign of either $\mathbf{e}$ or $\mathbf{c}$. This yields the following 'parametric' version of Moser's theorem.

Proposition 8.17 *The stereographic projection $\Psi\colon S^3 \to \mathbb{R}^3 \cup \{\infty\}$ sends the great circle $\boldsymbol{\gamma}_\alpha$ to the velocity curve of the Kepler solution with $h = -1/2$,*

$$\mathbf{e} = \sin\alpha\,(1, 0, 0) \quad \textit{and} \quad \mathbf{c} = \mu\cos\alpha\,(0, 0, 1).$$

The arc length parametrisation of $\boldsymbol{\gamma}_\alpha$ corresponds to the parametrisation of the velocity curve by the eccentric anomaly. □

It is instructive to refer back to Figure 8.9 and the discussion following it for the kind of phenomena that occur when either $\mathbf{e}$ or $\mathbf{c}$ passes through $\mathbf{0}$.

Moser's theorem confirms that the regularisation of collision solutions by turning them into 'bounce' orbits as in Exercise 4.7 is the right thing to do. When we turn from the Kepler solutions (for $h = -1/2$) to their hodographs, viewed under Ψ^{-1} as curves on S^3, there is no longer anything exceptional about these bounce orbits: every Kepler solution corresponds to a unique unit speed geodesic; the collision solutions are those where the corresponding great circle passes through N. Moreover, the parametrisation in terms of the eccentric anomaly, which initially may have looked artificial, now receives a geometric justification.

8.4 Hyperbolic geometry

We now want to develop an analogous theory for solutions of the Kepler problem with positive energy. In the negative energy case, Moser's theorem relates the velocity circles to geodesics on the 3-sphere. For positive energy, the velocity circles will be seen to correspond to geodesics in what is called hyperbolic 3-space. In this section I give a brief introduction to some of the basic concepts in hyperbolic geometry.

Consider the open half-space

$$\mathbb{H} := \{\mathbf{x} = (x_1, \ldots, x_n) \in \mathbb{R}^n : \ x_n > 0\}.$$

Let $\boldsymbol{\gamma}: [a,b] \to \mathbb{H}$ be a C^1-curve. We shall assume that $\boldsymbol{\gamma}$ is **regular**, i.e. its velocity vector is nowhere zero. Write $\boldsymbol{\gamma} = (\gamma_1, \ldots, \gamma_n)$.

Definition 8.18 The **hyperbolic length** of $\boldsymbol{\gamma}$ is

$$\mathcal{L}_{\mathrm{h}}(\boldsymbol{\gamma}) := \int_a^b \frac{|\dot{\boldsymbol{\gamma}}(t)|}{\gamma_n(t)}\, \mathrm{d}t.$$

The hyperbolic length $\mathcal{L}_{\mathrm{h}}(\boldsymbol{\gamma})$ depends only on the trace $\boldsymbol{\gamma}([a,b]) \subset \mathbb{H}$ of the regular curve $\boldsymbol{\gamma}$, not on the specific (regular) parametrisation of the curve.[6]

Definition 8.19 A **hyperbolic geodesic** is a curve $\boldsymbol{\gamma}: I \to \mathbb{H}$, defined on some interval $I \subset \mathbb{R}$, with the following properties.

(i) The parametrisation is proportional to arc length, i.e. the function $|\dot{\boldsymbol{\gamma}}|/\gamma_n$ is constant.

(ii) The curve is locally distance minimising, i.e. for sufficiently small subintervals $[t_0, t_1] \subset I$, the curve $\boldsymbol{\gamma}|_{[t_0,t_1]}$ is the shortest connection from $\boldsymbol{\gamma}(t_0)$ to $\boldsymbol{\gamma}(t_1)$.[7]

[6] Given a diffeomorphism $\varphi: [c,d] \to [a,b]$, the fact that $\mathcal{L}_{\mathrm{h}}(\gamma \circ \varphi) = \mathcal{L}_{\mathrm{h}}(\gamma)$ is a simple application of the transformation formula for integrals. To show that two regular C^1-curves with the same trace differ by such a reparametrisation requires an application of the implicit function theorem.

[7] The proviso 'sufficiently small' will turn out to be superfluous for curves in $\mathbb{H}$, but, in general,

Example 8.20 The unique hyperbolic geodesic γ (of unit speed) joining two points

$$(a_1, \ldots, a_{n-1}, \mathrm{e}^a) \quad \text{and} \quad (a_1, \ldots, a_{n-1}, \mathrm{e}^b)$$

in $\mathbb{H}$ on the same vertical line (with $a < b$, say) is the arc of that vertical line, parametrised by

$$\gamma(s) = (a_1, \ldots, a_{n-1}, \mathrm{e}^s), \quad s \in [a, b].$$

Indeed, we have $|\gamma'(s)|/\gamma_n(s) = 1$ and hence $\mathcal{L}_\mathrm{h}(\gamma) = b - a$. Now let

$$[0, 1] \ni t \longmapsto \boldsymbol{\beta}(t) \in \mathbb{H}$$

be a regular C^1-curve joining $\boldsymbol{\beta}(0) = \gamma(a)$ with $\boldsymbol{\beta}(1) = \gamma(b)$. We have

$$\frac{|\dot{\boldsymbol{\beta}}(t)|}{\beta_n(t)} \geq \frac{\dot{\beta}_n(t)}{\beta_n(t)},$$

with equality if and only if $\dot{\beta}_1(t) = \ldots = \dot{\beta}_{n-1}(t) = 0$ and $\dot{\beta}_n(t) \geq 0$; regularity then forces $\dot{\beta}_n(t) > 0$. Hence

$$\mathcal{L}_\mathrm{h}(\boldsymbol{\beta}) \geq \int_0^1 \frac{\dot{\beta}_n(t)}{\beta_n(t)}\, \mathrm{d}t = \log \beta_n(t)\Big|_{t=0}^{t=1} = \log \mathrm{e}^b - \log \mathrm{e}^a = b - a,$$

with equality if and only if $\boldsymbol{\beta}$ is a regular parametrisation of the straight line segment joining $\gamma(a)$ with $\gamma(b)$.

Definition 8.21 The **tangent space** $T_\mathbf{x}\mathbb{H}$ of $\mathbb{H}$ at a point $\mathbf{x} \in \mathbb{H}$ is the space of velocity vectors of C^1-curves passing through $\mathbf{x}$:

$$T_\mathbf{x}\mathbb{H} := \{\dot{\gamma}(0), \text{ where } \gamma\colon (-\varepsilon, \varepsilon) \to \mathbb{H} \text{ is a } C^1\text{-curve with } \gamma(0) = \mathbf{x}\}.$$

This space can be naturally identified with $\mathbb{R}^n$, since $\mathbf{v} \in \mathbb{R}^n$ can be thought of as the velocity vector of the curve $t \mapsto \mathbf{x} + t\mathbf{v}$. On $T_\mathbf{x}\mathbb{H}$ we have the scalar product[8]

$$\boxed{\langle \mathbf{v}, \mathbf{w} \rangle_\mathrm{h} := \frac{\langle \mathbf{v}, \mathbf{w} \rangle}{x_n^2},}$$

where $\langle\, .\,, .\, \rangle$ denotes the standard euclidean inner product on $\mathbb{R}^n$. Such a pointwise inner product on tangent spaces, depending smoothly on the point $\mathbf{x}$, is called a **Riemannian metric**. The associated norm is given by

$$|\mathbf{v}|_\mathrm{h} := \sqrt{\langle \mathbf{v}, \mathbf{v} \rangle_\mathrm{h}}.$$

it is an essential part of the definition of geodesics. For instance, in spherical geometry an arc of a great circle longer than the distance between two antipodal points is locally distance minimising, but not globally so.

[8] Strictly speaking, we should include the point $\mathbf{x}$ in the notation, but this point will always be clear from the context.

With this notation, the hyperbolic length of a regular C^1-curve $\gamma\colon\ [a,b] \to \mathbb{H}$ can be written as $\mathcal{L}_\mathrm{h}(\gamma) = \int_a^b |\dot{\gamma}(t)|_\mathrm{h}\,\mathrm{d}t$.

Definition 8.22 The pair $(\mathbb{H}, \langle\,.\,,.\,\rangle_\mathrm{h})$ consisting of the half-space $\mathbb{H}$ and the Riemannian metric $\langle\,.\,,.\,\rangle_\mathrm{h}$ is called **the half-space model of hyperbolic space**.

In what follows the Riemannian metric will be understood, and we simply write $\mathbb{H}$ for hyperbolic space.

Definition 8.23 Given a C^1-map $\Phi\colon\ \mathbb{H} \to \mathbb{H}$, its **differential** at the point $\mathbf{x} \in \mathbb{H}$ is the linear map

$$\begin{array}{ccc} T_\mathbf{x}\mathbb{H} & \longrightarrow & T_{\Phi(\mathbf{x})}\mathbb{H} \\ \mathbf{v} & \longmapsto & J_{\Phi,\mathbf{x}}(\mathbf{v}) \end{array}$$

given by the Jacobian matrix $J_{\Phi,\mathbf{x}}$ of Φ at $\mathbf{x}$.

If γ is a C^1-curve in $\mathbb{H}$ through the point $\mathbf{x} = \gamma(0)$, the chain rule says

$$\frac{\mathrm{d}}{\mathrm{d}t}(\Phi \circ \gamma)(0) = J_{\Phi,\mathbf{x}}(\dot{\gamma}(0)).$$

In other words, geometrically the differential at $\mathbf{x}$ can be thought of as the linear map that sends the velocity vector at $\mathbf{x}$ of a curve γ to the velocity vector of the image curve $\Phi \circ \gamma$ at the point $\Phi(\mathbf{x})$.

Definition 8.24 An **isometry** of $\mathbb{H}$ is a diffeomorphism $\Phi\colon\ \mathbb{H} \to \mathbb{H}$ whose differential preserves the Riemannian metric, that is,

$$\langle J_{\Phi,\mathbf{x}}(\mathbf{v}), J_{\Phi,\mathbf{x}}(\mathbf{w})\rangle_\mathrm{h} = \langle \mathbf{v}, \mathbf{w}\rangle_\mathrm{h}$$

for all $\mathbf{x} \in \mathbb{H}$ and all $\mathbf{v}, \mathbf{w} \in T_\mathbf{x}\mathbb{H}$.

Beware that $J_{\Phi,\mathbf{x}}(\mathbf{v}) \in T_{\Phi(\mathbf{x})}\mathbb{H}$, so the inner product on the left-hand side has to be taken at the point $\Phi(\mathbf{x})$.

If Φ is an isometry of $\mathbb{H}$, then in particular it preserves the lengths of curves: $\mathcal{L}_\mathrm{h}(\Phi \circ \gamma) = \mathcal{L}_\mathrm{h}(\gamma)$; this is a consequence of the chain rule stated above.[9] It follows that an isometry sends geodesics to geodesics. By the same reasoning, isometries preserves angles, which are defined as follows.

Definition 8.25 The **angle** $\measuredangle(\mathbf{v}, \mathbf{w}) \in [0, \pi]$ between two tangent vectors $\mathbf{v}, \mathbf{w} \in T_\mathbf{x}\mathbb{H}$ is determined by

$$\cos \measuredangle(\mathbf{v}, \mathbf{w}) := \frac{\langle \mathbf{v}, \mathbf{w}\rangle_\mathrm{h}}{|\mathbf{v}|_\mathrm{h} \cdot |\mathbf{w}|_\mathrm{h}}.$$

[9] Thanks to the polarisation identity, see (8.7) below, the converse is also true: a diffeomorphism that preserves the lengths of curves is an isometry.

Observe that both numerator and denominator in this definition differ from the definition of euclidean angles by a factor $1/x_n^2$. So this factor cancels out, and hyperbolic angles are in fact the same as euclidean ones.

Remark 8.26 Two Riemannian metrics that differ by multiplication with a smooth positive function define the same notion of angle. Such metrics are called **conformally equivalent**. So the hyperbolic metric on $\mathbb{H}$ is conformally equivalent to the euclidean one.

As a final piece of notation we write

$$\partial\mathbb{H} := \{\mathbf{x} \in \mathbb{R}^n : \ x_n = 0\} \cup \{\infty\}$$

for the **boundary at infinity** of hyperbolic space $\mathbb{H}$. The reason for the terminology 'at infinity' is provided by Example 8.20. The geodesic line segment $s \mapsto (a_1, \ldots, a_{n-1}, \mathrm{e}^s)$, $s \in (-\infty, b]$ joining (asymptotically) the point $(a_1, \ldots, a_{n-1}, 0) \in \partial\mathbb{H}$ with $(a_1, \ldots, a_{n-1}, \mathrm{e}^b) \in \mathbb{H}$ has infinite length.

We now want to see that certain inversions constitute examples of hyperbolic isometries.

Lemma 8.27 *Let $\mathbf{p} \neq \infty$ be a point in $\partial\mathbb{H}$. The inversion Φ of $\mathbb{R}^n \cup \{\infty\}$ in a sphere centred at $\mathbf{p}$ restricts to an isometry of $\mathbb{H}$.*

Proof The inversion Φ clearly sends $\mathbb{H}$ to itself, so we need only investigate its metric properties. First we want to show that without loss of generality we may assume $\mathbf{p} = \mathbf{0}$. For any 'horizontal' vector $\mathbf{c} = (c_1, \ldots, c_{n-1}, 0)$, the translation $T_\mathbf{c}\colon \mathbf{x} \mapsto \mathbf{x} + \mathbf{c}$ of $\mathbb{H}$ is an isometry, since $\frac{\mathrm{d}}{\mathrm{d}t}(T_\mathbf{c} \circ \gamma)(t) = \dot{\gamma}(t)$ and $(T_\mathbf{c} \circ \gamma)_n(t) = \gamma_n(t)$. If $\Phi_\mathbf{p}$ denotes inversion in the sphere of radius r centred at $\mathbf{p} \in \partial\mathbb{H} \setminus \{\infty\}$, then $T_\mathbf{p}^{-1} \circ \Phi_\mathbf{p} \circ T_\mathbf{p} = \Phi_\mathbf{0}$.

So we may indeed assume $\mathbf{p} = \mathbf{0}$. Let $t \mapsto \mathbf{x}(t)$ be a C^1-curve in $\mathbb{H}$, and let $\mathbf{y} = r^2\mathbf{x}/|\mathbf{x}|^2$ be its image under the inversion in a sphere of radius r centred at $\mathbf{0}$. By Definition 8.23, the image of $\dot{\mathbf{x}}(t)$ under the differential of the inversion at the point $\mathbf{x}(t)$ is the velocity vector $\dot{\mathbf{y}}(t)$ of the image curve.

We compute

$$\dot{\mathbf{y}} = r^2 \frac{\dot{\mathbf{x}}}{|\mathbf{x}|^2} - r^2\mathbf{x}\,\frac{2\langle\dot{\mathbf{x}}, \mathbf{x}\rangle}{|\mathbf{x}|^4}.$$

Hence

$$\begin{aligned}
|\dot{\mathbf{y}}|^2 &= \langle\dot{\mathbf{y}}, \dot{\mathbf{y}}\rangle \\
&= r^4 \frac{\langle\dot{\mathbf{x}}, \dot{\mathbf{x}}\rangle}{|\mathbf{x}|^4} - 4\frac{r^4}{|\mathbf{x}|^6}\langle\dot{\mathbf{x}}, \mathbf{x}\rangle^2 + 4r^4\frac{\langle\dot{\mathbf{x}}, \mathbf{x}\rangle^2}{|\mathbf{x}|^8}\langle\mathbf{x}, \mathbf{x}\rangle \\
&= r^4 \frac{|\dot{\mathbf{x}}|^2}{|\mathbf{x}|^4},
\end{aligned}$$

or $|\dot{\mathbf{y}}| = r^2|\dot{\mathbf{x}}|/|\mathbf{x}|^2$. It follows that

$$\frac{|\dot{\mathbf{y}}|}{|\mathbf{y}|} = r^2 \frac{|\dot{\mathbf{x}}|}{|\mathbf{x}|^2} \cdot \frac{|\mathbf{x}|}{r^2} = \frac{|\dot{\mathbf{x}}|}{|\mathbf{x}|}.$$

Since $\mathbf{x}$ and $\mathbf{y}$ (at any given time t) lie on a radial line, we have $|\mathbf{x}|/|\mathbf{y}| = x_n/y_n$. We conclude

$$|\dot{\mathbf{y}}|_{\mathrm{h}} = \frac{|\dot{\mathbf{y}}|}{y_n} = \frac{|\dot{\mathbf{x}}|}{x_n} = |\dot{\mathbf{x}}|_{\mathrm{h}},$$

so the differential of Φ preserves the hyperbolic norm of tangent vectors. By the polarisation identity[10]

$$\langle \mathbf{v}, \mathbf{w} \rangle_{\mathrm{h}} = \frac{1}{2}\left(|\mathbf{v}+\mathbf{w}|_{\mathrm{h}}^2 - |\mathbf{v}|_{\mathrm{h}}^2 - |\mathbf{w}|_{\mathrm{h}}^2\right), \tag{8.7}$$

angles between tangent vectors are being preserved, too. □

Definition 8.28 A **hyperbolic line** is a maximal geodesic (understood to be of unit speed) in $\mathbb{H}$, i.e. a geodesic $\gamma\colon \mathbb{R} \to \mathbb{H}$ defined on all of $\mathbb{R}$.

Up to a constant shift in the parameter, a hyperbolic line γ is determined by its trace $\gamma(\mathbb{R})$, so I shall also refer to that trace as the 'hyperbolic line'.

Proposition 8.29 *The hyperbolic lines are precisely the euclidean half-lines orthogonal to $\partial\mathbb{H}$ and the euclidean semicircles orthogonal to $\partial\mathbb{H}$.*

Proof By Example 8.20, the euclidean half-line

$$s \longmapsto (a_1, \ldots, a_{n-1}, \mathrm{e}^s), \quad s \in \mathbb{R},$$

is a maximal geodesic.

Now let $C \subset \mathbb{H}$ be a semicircle orthogonal to $\partial\mathbb{H}$ at the points $\mathbf{p}, \mathbf{q}$. Let Φ be the inversion of $\mathbb{R}^n \cup \{\infty\}$ in a sphere centred at $\mathbf{p}$. By Lemma 8.27 this restricts to an isometry of $\mathbb{H}$. By Proposition 8.10, Φ sends C to a half-line, which has to be orthogonal to $\partial\mathbb{H}$ because of the symmetry with respect to $\partial\mathbb{H}$, see Figure 8.10. Notice that the point of intersection between C and the sphere of inversion stays fixed under Φ; this determines the line $\Phi(C)$. Thus, we have established that such semicircles are likewise hyperbolic lines.

In order to show that there are no other types of hyperbolic lines, we first observe that through any two distinct points $\mathbf{x}, \mathbf{y} \in \mathbb{H}$ there is a unique hyperbolic line of one of the two types just described. For $\mathbf{x}$ and $\mathbf{y}$ on a vertical half-line this is clear. If $\mathbf{x}$ and $\mathbf{y}$ do not lie on a vertical half-line, they lie on a unique semicircle C orthogonal to $\partial\mathbb{H}$: draw the perpendicular bisector of the euclidean line segment $\mathbf{xy}$ in the plane orthogonal to $\partial\mathbb{H}$ determined by

[10] This identity, as one can easily show, holds for any scalar product and the associated norm.

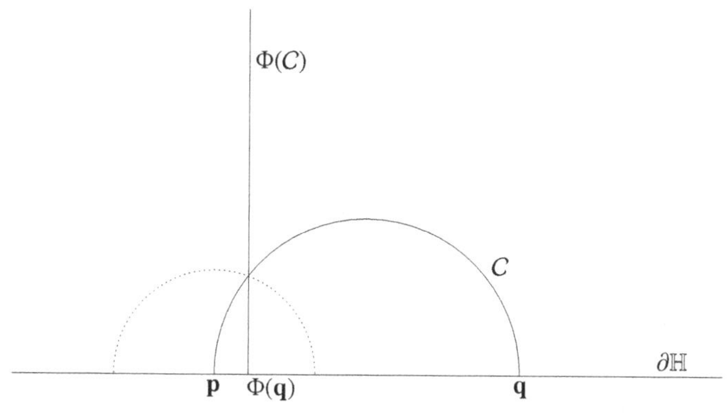

Figure 8.10 Hyperbolic lines.

these two points; the intersection of this bisector with $\partial\mathbb{H}$ is the centre of C, see Figure 8.11.

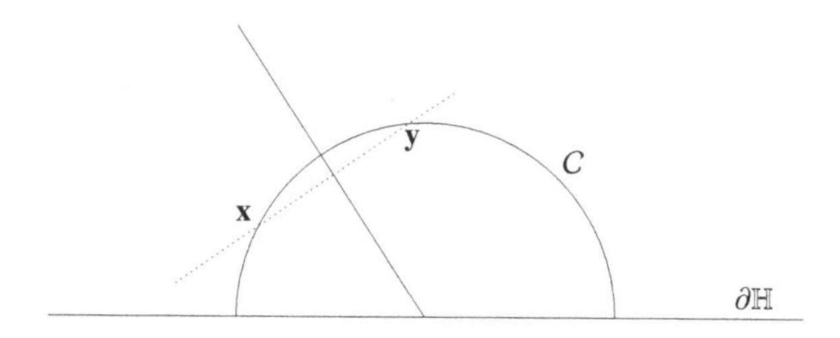

Figure 8.11 Finding a hyperbolic line C through $\mathbf{x}$ and $\mathbf{y}$.

If $\mathbf{x}$ and $\mathbf{y}$ lie on a half-line, this half-line is the unique hyperbolic geodesic containing $\mathbf{x}$ and $\mathbf{y}$ by Example 8.20. If $\mathbf{x}$ and $\mathbf{y}$ lie on a semicircle, then, as above, one can find a hyperbolic isometry sending this semicircle to a half-line. Since isometries send geodesics to geodesics, this again implies uniqueness of the hyperbolic line containing $\mathbf{x}$ and $\mathbf{y}$. □

As we have just seen, in the half-space model of hyperbolic space it is quite easy to determine all geodesics. For the purpose of proving an analogue of Moser's theorem, however, we shall have to work with the Poincaré disc model of hyperbolic space, which we introduce next.

The idea is to transform $\mathbb{H}$ to an n-dimensional disc $\mathbb{D}$ by a suitable inversion of $\mathbb{R}^n$ and to equip $\mathbb{D}$ with the induced metric. Then, by construction, the map $\mathbb{H} \to \mathbb{D}$ becomes an isometry, and from a metric point of view the two spaces are indistinguishable.

Let $\Phi\colon \mathbb{R}^n \cup \{\infty\} \to \mathbb{R}^n \cup \{\infty\}$ be the inversion in the sphere of radius

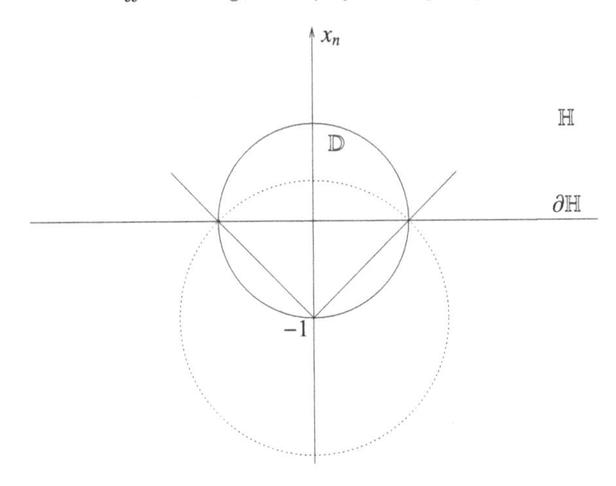

Figure 8.12 The inversion sending $\mathbb{H}$ to $\mathbb{D}$.

$\sqrt{2}$ about the point $(0, \ldots, 0, -1)$, see Figure 8.12. By Proposition 8.10, the inversion Φ sends the $(n-1)$-plane $\partial\mathbb{H}$ to an $(n-1)$-sphere. We observe that Φ sends the point $\mathbf{0} \in \partial\mathbb{H}$ to $(0, \ldots, 0, 1)$, and it fixes the $(n-2)$-sphere in $\partial\mathbb{H}$ of radius 1 about $\mathbf{0}$. This implies that $\Phi(\partial\mathbb{H})$ is the $(n-1)$-sphere of radius 1 about $\mathbf{0}$. So $\Phi(\mathbb{H})$ has to be equal either to the open n-disc $\mathbb{D}$ of radius 1 about $\mathbf{0}$ or to the complement of the closure of $\mathbb{D}$.

In order to decide which alternative holds, it suffices to consider a single point in $\mathbb{H}$. The point $(0, \ldots, 0, 1) \in \mathbb{H}$ is sent to the point $\mathbf{0} \in \mathbb{D}$; this implies $\Phi(\mathbb{H}) = \mathbb{D}$.

Proposition 8.30 *The Riemannian metric on $\mathbb{D}$ induced from the hyperbolic metric on $\mathbb{H}$ via the diffeomorphism $\Phi\colon \mathbb{H} \to \mathbb{D}$ is given by*

$$\boxed{\langle \mathbf{v}, \mathbf{w} \rangle_{\mathrm{h}'} = \frac{4\langle \mathbf{v}, \mathbf{w} \rangle}{(1 - |\mathbf{y}|^2)^2} \quad \textit{for } \mathbf{v}, \mathbf{w} \in T_{\mathbf{y}}\mathbb{D}.}$$

Proof Consider a C^1-curve $t \mapsto \mathbf{x}(t)$ in $\mathbb{H}$ and its image $t \mapsto \mathbf{y}(t) := \Phi(\mathbf{x}(t))$ in $\mathbb{D}$. With $\mathbf{u} := (0, \ldots, 0, 1)$, the relation between $\mathbf{x}$ and $\mathbf{y}$ can be written as

$$\mathbf{x} + \mathbf{u} = \frac{2(\mathbf{y} + \mathbf{u})}{|\mathbf{y} + \mathbf{u}|^2}, \tag{8.8}$$

since Φ was defined as the inversion in the sphere of radius $\sqrt{2}$ about the point $-\mathbf{u}$. Notice that $\langle \mathbf{x}, \mathbf{u} \rangle = x_n$, so the hyperbolic length of $\dot{\mathbf{x}}$ can be written as $|\dot{\mathbf{x}}|_{\mathrm{h}} = |\dot{\mathbf{x}}| / \langle \mathbf{x}, \mathbf{u} \rangle$.

The induced norm on the tangent spaces to $\mathbb{D}$ is given by

$$|\dot{\mathbf{y}}|_{h'} := |\dot{\mathbf{x}}|_h.$$

We compute

$$\dot{\mathbf{x}} = \frac{2\dot{\mathbf{y}}|\mathbf{y}+\mathbf{u}|^2 - 2(\mathbf{y}+\mathbf{u})\cdot 2\langle \dot{\mathbf{y}}, \mathbf{y}+\mathbf{u}\rangle}{|\mathbf{y}+\mathbf{u}|^4},$$

hence

$$\begin{aligned}
|\dot{\mathbf{x}}|^2 &= \frac{1}{|\mathbf{y}+\mathbf{u}|^8}\left(4|\dot{\mathbf{y}}|^2|\mathbf{y}+\mathbf{u}|^4 - 16\langle \dot{\mathbf{y}}, \mathbf{y}+\mathbf{u}\rangle^2|\mathbf{y}+\mathbf{u}|^2 + 16|\mathbf{y}+\mathbf{u}|^2\langle \dot{\mathbf{y}}, \mathbf{y}+\mathbf{u}\rangle^2\right)\\
&= \frac{4|\dot{\mathbf{y}}|^2}{|\mathbf{y}+\mathbf{u}|^4}.
\end{aligned}$$

It follows that

$$|\dot{\mathbf{x}}| = \frac{2|\dot{\mathbf{y}}|}{|\mathbf{y}+\mathbf{u}|^2}.$$

Moreover, from (8.8) we have

$$\langle \mathbf{x}, \mathbf{u}\rangle = \frac{2(\langle \mathbf{y}, \mathbf{u}\rangle + 1)}{|\mathbf{y}+\mathbf{u}|^2} - 1,$$

and because of

$$|\mathbf{y}+\mathbf{u}|^2 = |\mathbf{y}|^2 + 2\langle \mathbf{y}, \mathbf{u}\rangle + 1$$

this can be written as

$$\langle \mathbf{x}, \mathbf{u}\rangle = \frac{1-|\mathbf{y}|^2}{|\mathbf{y}+\mathbf{u}|^2}.$$

We conclude that

$$|\dot{\mathbf{y}}|_{h'} = |\dot{\mathbf{x}}|_h = \frac{2|\dot{\mathbf{y}}|}{1-|\mathbf{y}|^2}.$$

By polarisation we obtain the claimed Riemannian metric. □

Definition 8.31 The pair $(\mathbb{D}, \langle\,.\,,\,.\,\rangle_{h'})$ is called **the Poincaré disc model of hyperbolic space**.

Proposition 8.32 *The hyperbolic lines in the Poincaré disc model are precisely the arcs of circles orthogonal to $\partial\mathbb{D}$ and the diameters of $\mathbb{D}$.*

Proof By construction of the metric on $\mathbb{D}$, the diffeomorphism $\Phi\colon \mathbb{H} \to \mathbb{D}$ is an isometry, and so the hyperbolic lines in $\mathbb{D}$ are precisely the images under Φ of the hyperbolic lines in $\mathbb{H}$. By Proposition 8.29, the hyperbolic lines in $\mathbb{H}$ are given by euclidean lines or circles orthogonal to $\partial\mathbb{H}$. By Proposition 8.10, the inversion Φ maps these to lines or circles orthogonal to $\partial\mathbb{D}$. □

8.5 The theorem of Osipov and Belbruno

Moser's theorem says that the hodographs of Kepler solutions in $\mathbb{R}^3$ with energy $h = -1/2$ correspond, under stereographic projection, to the great circles on the 3-sphere.

Here is the analogue of this result for the hyperbolic case with energy $h = 1/2$ and the parabolic case, where $h = 0$. As in the elliptic case, the following statement is meant to include the collision orbits ($c = 0$), which are regularised to become bounce orbits. Recall that the hodograph is always regarded as an oriented curve.

Theorem 8.33 (Osipov, Belbruno) *The inversion in the unit sphere $S^2 \subset \mathbb{R}^3 \cup \{\infty\}$ yields a bijection between, on the one hand, the hodographs of Kepler solutions with energy $h = 1/2$ or $h = 0$ and, on the other hand, the oriented hyperbolic lines in $\mathbb{D}^3$ or euclidean lines in $\mathbb{R}^3$, respectively.*[11]

Proof Any geodesic in $\mathbb{D}^3$ or $\mathbb{R}^3$, respectively, lies in a radial plane, see Exercise 8.15. Likewise, any velocity curve and its image under inversion lie in a radial plane, the plane of motion. By the rotational symmetry of the Kepler problem (Exercise 1.1), we may therefore restrict our attention to a single plane of motion, which we take to be the xy-plane. Correspondingly, we have to consider geodesics in $\mathbb{D}^2$ or $\mathbb{R}^2$, respectively.

In the case $h = 1/2$ and $c \neq 0$, we found in Proposition 8.5 and Lemma 8.6 that the velocity circles are precisely those circles that intersect the energy circle – which for $h = 1/2$ is the unit circle – orthogonally. The hodograph is the part of the velocity circle outside the unit disc. After inversion, we obtain all possible arcs of circles in $\mathbb{D}^2$ orthogonal to $\partial\mathbb{D}^2$. For $c = 0$, the hodograph is, by the energy formula (3.6), the part of a radial ray outside the unit disc. Speed 1 is reached asymptotically for $r \to \infty$. After regularisation of the collision (where the velocity becomes infinite and then changes sign), the hodograph is the part of a radial line outside the unit disc, including the point ∞. After inversion, this becomes a diameter of $\mathbb{D}^2$. By Proposition 8.32, this exhausts all hyperbolic lines in $\mathbb{D}^2$.

For $h = 0$ and $c \neq 0$, the hodographs are circles through the origin, with that origin removed; for $c = 0$ and after regularisation, the hodographs are radial lines without the origin, but including the point ∞. After inversion in the unit circle, we obtain all (euclidean) lines in $\mathbb{R}^2$. □

Next we come to the parametric version of Theorem 8.33, i.e. the analogue of Proposition 8.17. We begin with the parabolic case. Parametrise the velocity

[11] The geodesics in euclidean space are simply the straight lines. This can be proved by reasoning as in Example 8.20.

curves in terms of the scaled eccentric anomaly $s := u/\sqrt{\mu}$. We shall see that this parameter s permits a natural definition also in the case $c = 0$. The reasons for this particular choice of parameter and why we scale the metric in the following statement will become clear at the end of this section, where we discuss the result for arbitrary energies h.

Proposition 8.34 *Let $s \mapsto \mathbf{v}(s)$ be the velocity curve of a Kepler solution with $h = 0$. The inversion in the unit sphere sends this curve to a euclidean line parametrised by arc length with respect to the scaled euclidean metric $4\langle\,.\,,.\,\rangle$.*

Proof The rotational symmetry of the Kepler problem allows us to assume that the orbit lies in the xy-plane and $\mathbf{c} = (0, 0, c)$, i.e. for $c \neq 0$ the orbit is traversed in the counter-clockwise direction. In addition, we may restrict our attention to solutions with eccentricity vector pointing in the positive x-direction.

For $c \neq 0$ the velocity curve is then given by (8.1) with $e = 1$:

$$\mathbf{v}(\theta) = \frac{\mu}{c}\,(-\sin\theta, 1 + \cos\theta), \quad \theta \in (-\pi, \pi).$$

This gives

$$v^2 = \frac{2\mu^2}{c^2}\,(1 + \cos\theta).$$

Let $\mathbf{w}$ be the curve obtained by inverting $\mathbf{v}$ in the unit circle:

$$\mathbf{w}(\theta) = \frac{\mathbf{v}(\theta)}{v^2(\theta)} = \frac{c}{2\mu}\left(\frac{-\sin\theta}{1 + \cos\theta}, 1\right).$$

The parametrisations of the Kepler solution $\mathbf{r}$ in terms of the true anomaly θ and the eccentric anomaly u, respectively, are given by

$$r(\theta)\cdot(\cos\theta, \sin\theta) = \mathbf{r} = \left(\frac{1}{2}\left(\frac{c^2}{\mu} - u^2\right), \frac{c}{\sqrt{\mu}}u\right),$$

see Section 4.3. We then compute

$$\frac{c}{2\mu}\cdot\frac{-\sin\theta}{1 + \cos\theta} = \frac{-\sin\theta}{2c}\cdot\frac{c^2/\mu}{1 + \cos\theta} = \frac{-r\sin\theta}{2c} = -\frac{u}{2\sqrt{\mu}} = -\frac{s}{2},$$

i.e. $\mathbf{w}(s) = (-s/2, c/2\mu)$, which describes the arc length parametrisation of a horizontal line with respect to the metric $4\langle\,.\,,.\,\rangle$.

In the limit $c \to 0$, the curve $\mathbf{r}$ becomes

$$\mathbf{r}(u) = (-u^2/2, 0),$$

where by Proposition 4.7 the relation between u and t is given by $u^3/6 = \sqrt{\mu}\,t$.

It is straightforward to verify that for $u \neq 0$ this does indeed define a solution of (K). If we allow u to vary over all of $\mathbb{R}$, we have the description of the regularised solution, with $t = 0$ the time of collision.[12]

Write $x(t) = -u^2(t)/2$. The equation for u gives $u^2\dot{u} = 2\sqrt{\mu}$. The inverted velocity curve is given by $\mathbf{w} = (1/\dot{x}, 0)$ with

$$\dot{x} = -u\dot{u} = -2\sqrt{\mu}/u = -2/s,$$

hence $\mathbf{w} = (-s/2, 0)$, which is smooth even in $s = 0$.

In this fashion we realise all horizontal lines in the upper half of the xy-plane (including the x-axis), oriented in the negative x-direction. □

Remark 8.35 Given a solution $\mathbf{r}$ of (K) with $h = 0$, the parameter $s = u/\sqrt{\mu}$ can be recovered as

$$s(t) = \int_0^t \frac{\mathrm{d}\tau}{r(\tau)}, \tag{8.9}$$

where $t = 0$ is the time of pericentre passage or collision, respectively. For $c \neq 0$ this is the content of Exercise 4.6. For the case $c = 0$ see Exercise 8.17.

In the case $h = 1/2$ there is no need to rescale the eccentric anomaly u (see Remark 8.41 below). Again we shall see that this parameter has a natural extension to Kepler solutions with $c = 0$.

Proposition 8.36 *Let $u \mapsto \mathbf{v}(u)$ be the velocity curve of a Kepler solution with $h = 1/2$. The inversion in the unit sphere sends this curve to a hyperbolic line parametrised by arc length.*

Proof As in the case $h = 0$ we need only consider solutions in the xy-plane with angular momentum vector pointing in the positive z-direction (or equal to $\mathbf{0}$) and eccentricity vector pointing in the positive x-direction.

For $c \neq 0$ the velocity curve is given by (8.1) with $e > 1$ and, by (3.7), $c = \mu\sqrt{e^2 - 1}$:

$$\mathbf{v}(\theta) = \frac{1}{\sqrt{e^2 - 1}}(-\sin\theta, e + \cos\theta),$$

where the range of θ is determined by $1 + e\cos\theta > 0$. Straightforward compu-

[12] Beware that, as in the elliptic case, the functions $u \mapsto r(u)$ and $u \mapsto t(u)$ are smooth on $\mathbb{R}$, but the functions $t \mapsto r(t)$ and $t \mapsto u(t)$ are differentiable only on $\mathbb{R} \setminus \{0\}$.

tations yield the following formulae:

$$\begin{aligned}
v^2 &= \frac{1+e^2+2e\cos\theta}{e^2-1},\\
\mathbf{w} := \frac{\mathbf{v}}{v^2} &= \frac{\sqrt{e^2-1}}{1+e^2+2e\cos\theta}(-\sin\theta, e+\cos\theta),\\
\left|\frac{\mathrm{d}\mathbf{w}}{\mathrm{d}\theta}\right| &= \frac{\sqrt{e^2-1}}{1+e^2+2e\cos\theta},\\
w^2 &= \frac{e^2-1}{1+e^2+2e\cos\theta},\\
1-w^2 &= \frac{2(1+e\cos\theta)}{1+e^2+2e\cos\theta} = \frac{2c^2/\mu}{r(1+e^2+2e\cos\theta)}.
\end{aligned}$$

Moreover, we have

$$\dot{u} = \sqrt{\frac{\mu}{a}}\cdot\frac{1}{r}; \tag{8.10}$$

this is proved in a manner analogous to (4.2). For $h = 1/2$ and $c \neq 0$ we have $\mu = a$ by (3.8). We then compute

$$\left|\frac{\mathrm{d}\mathbf{w}}{\mathrm{d}u}\right| = \left|\frac{\mathrm{d}\mathbf{w}}{\mathrm{d}\theta}\right|\cdot\left|\frac{\mathrm{d}\theta}{\mathrm{d}t}\right|\cdot\left|\frac{\mathrm{d}t}{\mathrm{d}u}\right| = \frac{\sqrt{e^2-1}}{1+e^2+2e\cos\theta}\cdot\frac{c}{r^2}\cdot r,$$

hence

$$\left|\frac{\mathrm{d}\mathbf{w}}{\mathrm{d}u}\right|_{\mathrm{h}'} = \frac{2\,|\mathrm{d}\mathbf{w}/\mathrm{d}u|}{1-w^2} = \sqrt{e^2-1}\cdot\frac{\mu}{c} = 1.$$

So $u \mapsto \mathbf{w}(u)$ is indeed a unit speed parametrisation.

The solution for $c = 0$ and $\mathbf{e} = (1, 0)$ is found as in Exercise 4.7 by describing r and t as functions of u and taking the limit $e \to 1$. With $\sqrt{\mu/a} = 1$ this yields

$$x(u) = a(1-\cosh u), \quad t(u) = a(\sinh u - u). \tag{8.11}$$

Hence

$$\dot{x} = \frac{\mathrm{d}x}{\mathrm{d}u}\cdot\frac{\mathrm{d}u}{\mathrm{d}t} = \frac{\mathrm{d}x}{\mathrm{d}u}\cdot\left(\frac{\mathrm{d}t}{\mathrm{d}u}\right)^{-1} = \frac{-\sinh u}{\cosh u - 1},$$

where the same caveat holds about $t = 0$ as in the case $h = -1/2$. The inverted velocity curve

$$u \longmapsto \mathbf{w}(u) = \left(\frac{1-\cosh u}{\sinh u}, 0\right)$$

is smooth on all of $\mathbb{R}$. A simple calculation shows that this is again a unit speed curve in terms of the hyperbolic metric on $\mathbb{D}^2$. Notice that $\lim_{u\to\pm\infty}\mathbf{w}(u) = (\mp 1, 0)$. □

Remark 8.37 For a solution $\mathbf{r}$ of (K) with $h = -1/2$, the parameter u can be recovered by the same integral formula (8.9) as s in the previous remark. For $c \neq 0$ this follows from (8.10); for $c = 0$ the improper integral can be computed explicitly, starting from (8.11).

We have now seen how to relate the velocity circles of the Kepler problem for energies $h = \pm 1/2$ or 0 to geodesics in the unit sphere, the euclidean plane or the hyperbolic plane, respectively. I close this section with a brief discussion of the general case.

For $K \in \mathbb{R}^+$ consider the sphere $S^3_{1/\sqrt{K}} \subset \mathbb{R}^4$ of radius $1/\sqrt{K}$ centred at $\mathbf{0}$. The stereographic projection $\Psi_K \colon S^3_{1/\sqrt{K}} \setminus \{N\} \to \mathbb{R}^3$, where $N = (0, 0, 0, 1/\sqrt{K})$, is given by

$$\Psi_K(x_1, x_2, x_3, x_4) = \frac{1}{1 - \sqrt{K}\, x_4}(x_1, x_2, x_3);$$

the inverse map is given by

$$\Psi_K^{-1}(w_1, w_2, w_3) = \frac{1}{Kw^2 + 1}\left(2w_1, 2w_2, 2w_3, \frac{Kw^2 - 1}{\sqrt{K}}\right),$$

where we write $\mathbf{w} = (w_1, w_2, w_3)$ and $w = |\mathbf{w}|$.

Now equip $S^3_{1/\sqrt{K}}$ with the Riemannian metric induced from $\mathbb{R}^4$, i.e. the inner product on each tangent space is simply defined as the restriction of the euclidean inner product $\langle\, .\, , .\, \rangle$ on $\mathbb{R}^4$.

Proposition 8.38 *The Riemannian metric $\langle\, .\, , .\, \rangle_K$ on $\mathbb{R}^3$ that turns Ψ_K into an isometry is given by*

$$\boxed{\langle\, .\, , .\, \rangle_K = \frac{4\langle\, .\, , .\, \rangle}{(1 + Kw^2)^2}.} \tag{8.12}$$

Proof As in the proof of Proposition 8.30 one considers a C^1-curve $t \mapsto \mathbf{w}(t)$ in $\mathbb{R}^3$ and its preimage $t \mapsto \mathbf{x}(t) := \Psi_K^{-1}(\mathbf{w}(t))$ in S^3. A straightforward computation gives

$$|\dot{\mathbf{w}}|_K^2 := |\dot{\mathbf{x}}|^2 = \frac{4|\dot{\mathbf{w}}|^2}{(1 + Kw^2)^2}.$$

The result follows by polarisation. □

We shall henceforth identify $S^3_{1/\sqrt{K}} \subset \mathbb{R}^4$ with $(\mathbb{R}^3 \cup \{\infty\}, \langle\, .\, , .\, \rangle_K)$, notably in the formulation of the theorem below.

For $K \in \mathbb{R}^-$ we can define a Riemannian metric on the disc $\mathbb{D}_{1/\sqrt{|K|}}$ of radius $1/\sqrt{|K|}$ by the same formula (8.12). For $K = 0$ this formula defines the scaled euclidean metric $4\langle\, .\, , .\, \rangle$ on $\mathbb{R}^3$.

Definition 8.39 The manifolds

$$M_K := \begin{cases} \mathbb{D}_{1/\sqrt{|K|}} & \text{for } K < 0 \\ \mathbb{R}^3 & \text{for } K = 0 \\ S^3_{1/\sqrt{K}} & \text{for } K > 0 \end{cases}$$

with the Riemannian metric $\langle\, .\, ,.\, \rangle_K$ are called the three-dimensional **space forms**.

The qualitative description of the geodesics in the M_K with $K < 0$ is the same as in the Poincaré disc model with $K = -1$; likewise, for $K > 0$ the geodesics are again the great circles. The space M_K can be characterised as the unique complete, simply connected three-dimensional Riemannian manifold of constant sectional curvature K, see the notes and references section.

The theorems of Moser and Osipov–Belbruno can now be combined into one beautiful statement covering all energies.

Definition 8.40 Let $\mathbf{r}$ be a solution of the Kepler problem with $t = 0$ the time of pericentre passage or of (regularised) collision with the force centre, respectively. The function

$$\boxed{s(t) := \int_0^t \frac{\mathrm{d}\tau}{r(\tau)}}$$

is called the **Levi-Civita parameter**.

Remark 8.41 Observe that, by Exercise 4.6 and equation (3.8), for $h = \pm 1/2$ and $c \neq 0$ the Levi-Civita parameter coincides with the eccentric anomaly. For other values of h, the two parameters differ by a multiplicative constant.

Theorem 8.42 (Moser, Osipov, Belbruno) *The inversion of $\mathbb{R}^3$ in the unit sphere S^2 gives a one-to-one correspondence between the velocity curves of Kepler solutions with energy h, parametrised by the Levi-Civita parameter, and the unit speed geodesics in the space form M_{-2h}.*

Proof Most of the work for proving this theorem has already been done. The fact that inversion in S^2 sets up a one-to-one correspondence between the velocity curves of Kepler solutions of energy h, on the one hand, and the geodesics in M_{-2h}, on the other, can be seen exactly as in the special cases $h \in \{0, \pm 1/2\}$.

In the case $h > 0$ we need only observe that the energy sphere $\{v^2 = 2h\}$ maps under inversion in S^2 to the sphere $\{w^2 = 1/2h\} = \partial\mathbb{D}_{1/\sqrt{2h}}$.

As in the proof of Lemma 8.12 one shows that the great circles on $S^3_{1/\sqrt{K}}$ map under Ψ_K to circles invariant under the negative inversion $\mathbf{w} \mapsto -\mathbf{w}/Kw^2$ in the sphere of radius $1/\sqrt{K}$. These are the circles in radial planes characterised by the condition $\langle \mathbf{w}_1, \mathbf{w}_2 \rangle = -1/K$ for any pair of intersection points $\mathbf{w}_1, \mathbf{w}_2$ with a radial line. Under inversion in S^2, we obtain circles characterised by the corresponding condition $\langle \mathbf{v}_1, \mathbf{v}_2 \rangle = -K$, which, by Lemma 8.6, are exactly the velocity circles of Kepler solutions with $2h = -K$.

It remains only to verify the claim about the respective parametrisations. The Kepler equation (K) can be written as

$$\dot{\mathbf{v}} = -\frac{\mu}{r^2} \cdot \frac{\mathbf{r}}{r}.$$

This gives $|\dot{\mathbf{v}}| = \mu/r^2$. Then

$$\left|\frac{\mathrm{d}\mathbf{v}}{\mathrm{d}s}\right| = |\dot{\mathbf{v}}| \cdot \frac{\mathrm{d}t}{\mathrm{d}s} = \frac{\mu}{r} = \frac{1}{2}v^2 - h,$$

where we have used the energy equation (3.6). It follows that the metric which makes $s \mapsto \mathbf{v}(s)$ a unit speed curve is given by

$$\frac{4\langle\,.\,,\,.\,\rangle}{(v^2 - 2h)^2}.$$

A computation as in the proof of Lemma 8.27 shows that, under inversion in the unit sphere, this metric is sent to

$$\frac{4\langle\,.\,,\,.\,\rangle}{(1 - 2hw^2)^2} = \langle\,.\,,\,.\,\rangle_{-2h}.$$

This completes the proof.[13] □

8.6 Projective geometry

Here I describe a proof of Kepler's first law based on a transformation of the plane $\mathbb{R}^2$ called polar reciprocation. As I explain towards the end of this section, this should really be regarded as a transformation of points and lines in the projective plane $\mathbb{RP}^2$. For our application, however, one needs to use the metric properties of polar reciprocation, which can be seen only in $\mathbb{R}^2$.

[13] The attentive reader will have noticed that we did without inversion in $S^2 \subset \mathbb{R}^3$ in Moser's theorem. The reason is that this inversion, as the last step in the proof has shown, is an isometry for the metric $\langle\,.\,,\,.\,\rangle_1$, and it preserves the set of circles invariant under negative inversion in the unit sphere.

8.6.1 Polar reciprocation

Polar reciprocation is a transformation of the plane $\mathbb{R}^2$ whose definition starts from the inversion in a circle as discussed in Section 8.2. We fix once and for all the circle C_0 of radius 1 about the origin $O \in \mathbb{R}^2$ as our circle of inversion. This choice is just a matter of convenience; changing the radius of the circle of inversion merely leads to a rescaling in the constructions that follow.

Definition 8.43 Let $A \in \mathbb{R}^2$ be a point different from O. The **polar** a of A is the line orthogonal to OA through the point A' inverse to A with respect to C_0.

Let $b \subset \mathbb{R}^2$ be a line not passing through O. Let B' be the foot of the perpendicular from O to b. The **pole** B of b is the inverse of B'.

From the definition it is clear that, if a is the polar of A, then A is the pole of a, and *vice versa*. If $A \in C_0$, then a is the tangent line to C_0 at A. Figure 8.13 shows how to construct pole and polar geometrically for $A \notin C_0$.

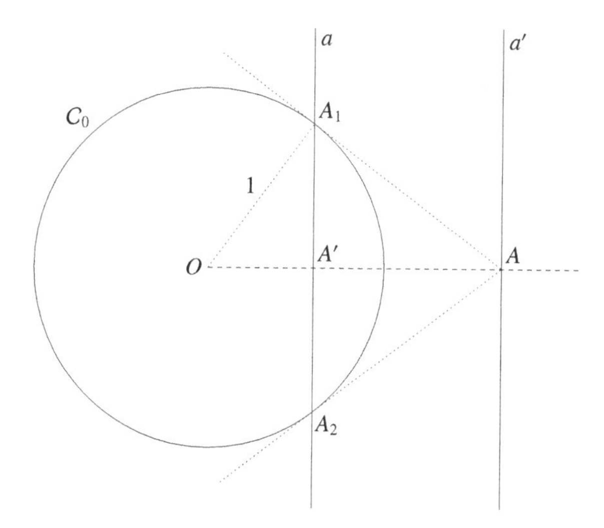

Figure 8.13 Construction of pole and polar.

Let A be a point in the exterior of C_0. The two tangents from A to C_0 have the respective points of contact A_1, A_2 with C_0. Let a be the line through A_1 and A_2, and A' the intersection of a with OA. Finally, let a' be the line through A orthogonal to OA.

Then $\triangle OAA_1$ is similar to $\triangle OA_1A'$, hence

$$OA \cdot OA' = OA_1 \cdot OA_1 = 1,$$

so A and A' are in inversion to each other. It follows that a is the polar of A, and a' the polar of A'.

Lemma 8.44 *Let A be the pole of a, and B the pole of b. If B lies on a, then A lies on b.*

Proof We start from a point A and its polar a. Given any point $B \in a$, let B' be the foot of the perpendicular from A to OB, see Figure 8.14. If B equals the inverse A' of A, then A lies on OB and coincides with B'. In this case the polar b of B passes through A by the very definition of poles and polars.

If $B \neq A'$, then $\triangle OBA'$ is a non-degenerate triangle similar to $\triangle OAB'$, hence

$$OB \cdot OB' = OA \cdot OA' = 1,$$

so B' is the inverse of B, and the line through A and B' is in fact the polar b of B. □

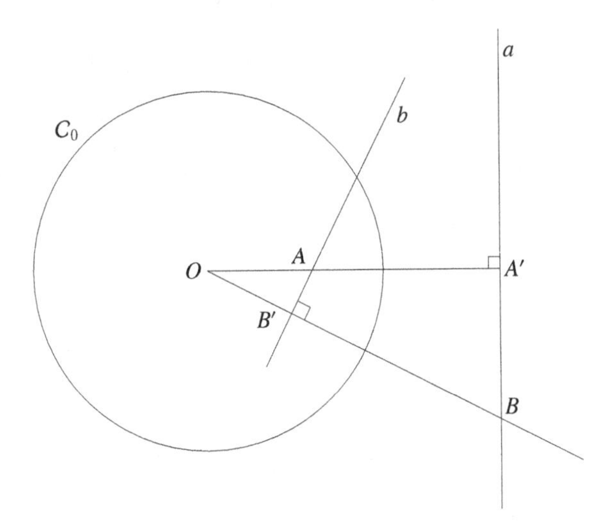

Figure 8.14 Duality between poles and polars.

Now let $\alpha\colon I \to \mathbb{R}^2$ be a C^1-curve. Write $\alpha(t) = (x(t), y(t))$. When we regard $\mathbb{R}^2$ as the xy-plane in $\mathbb{R}^3$, the angular momentum $\mathbf{c}$ about the origin $O \in \mathbb{R}^2$, as introduced in Definition 1.1, is then a vector pointing in the z-direction. In the present setting it is therefore appropriate to regard the angular momentum of α as the continuous function

$$\begin{array}{rccc} c\colon & I & \longrightarrow & \mathbb{R} \\ & t & \longmapsto & x(t)\dot{y}(t) - y(t)\dot{x}(t). \end{array}$$

The condition $c(t) \neq 0$ is equivalent to saying that $(x(t), y(t))$ and $(\dot{x}(t), \dot{y}(t))$ are linearly independent, or in other words that the tangent line through the point $(x(t), y(t))$ is defined and does not pass through O.

Definition 8.45 Let α be a C^1-curve in $\mathbb{R}^2$ with nowhere vanishing angular momentum. The **polar reciprocal** of α is the curve of poles of the tangent lines of α.

Proposition 8.46 *Let $t \mapsto \alpha(t) = (x(t), y(t))$, $t \in I$, be a C^1-curve with nowhere vanishing angular momentum c. Then the polar reciprocal α^* of α is given by*

$$\boxed{\alpha^*(t) = \frac{1}{c(t)}\,(\dot{y}(t), -\dot{x}(t)).}$$

Proof The tangent line $a = a_t$ to α at the point $\alpha(t) = (x(t), y(t))$ is parallel to the velocity vector $(\dot{x}(t), \dot{y}(t))$, and thus orthogonal to $(\dot{y}(t), -\dot{x}(t))$. Moreover, it contains the point $(x(t), y(t))$. Hence

$$a = \{(\xi, \eta) \in \mathbb{R}^2 : \ \xi\dot{y} - \eta\dot{x} = x\dot{y} - y\dot{x} = c\}.$$

So the foot A' of the perpendicular from O to a is

$$A' = \frac{c}{\dot{x}^2 + \dot{y}^2}\,(\dot{y}, -\dot{x}),$$

and the pole A of a is

$$A = \frac{1}{c}(\dot{y}, -\dot{x}).$$

□

8.6.2 Curvature of planar curves

As in Section 8.4 we say that a planar C^1-curve $t \mapsto \alpha(t) \in \mathbb{R}^2$, $t \in I$, is **regular** if its velocity vector $\dot{\alpha}(t)$ does not vanish for any $t \in I$. Then, for $t_0 \in I$,

$$s(t) := \int_{t_0}^{t} |\dot{\alpha}(t)|\,\mathrm{d}t$$

is the signed arc length of α measured from the point $\alpha(t_0)$. Because of

$$\frac{\mathrm{d}s}{\mathrm{d}t} = |\dot{\alpha}(t)| \neq 0,$$

the function $t \mapsto s(t)$ is invertible, so we may regard t as a function of s. The reparametrised curve

$$s \longmapsto \tilde{\alpha}(s) := \alpha(t(s))$$

is then a unit speed curve:

$$\tilde{\alpha}'(s) := \frac{\mathrm{d}\tilde{\alpha}}{\mathrm{d}s}(s) = \dot{\alpha}(t) \cdot \frac{\mathrm{d}t}{\mathrm{d}s} = \frac{\dot{\alpha}(t)}{|\dot{\alpha}(t)|}.$$

When $s \mapsto \alpha(s)$ is a unit speed C^2-curve, the norm $|\alpha''|$ of its acceleration is clearly a good measure for the 'bending' of the curve. For instance, for a line this acceleration will be zero. For a circle of radius r, the arc length parametrisation is given by

$$s \longmapsto r(\cos(s/r), \sin(s/r));$$

here the norm of the acceleration is $|\alpha''| = 1/r$.

If we also want to take into account whether the curve bends to the left or the right, we can regard $\mathbb{R}^2$ as $\mathbb{R}^2 \times \{0\} \subset \mathbb{R}^3$ and look at the vector $\alpha' \times \alpha''$, which points in the z-direction. Observe that the unit speed condition $\langle \alpha', \alpha' \rangle = 1$ implies $\langle \alpha', \alpha'' \rangle = 0$, so the acceleration α'' is orthogonal to the velocity α', and $|\alpha' \times \alpha''| = |\alpha''|$. If $\alpha' \times \alpha''$ points in the positive z-direction, the curve bends to the left.

If we write $\alpha'(s) = (\cos\theta(s), \sin\theta(s))$ with a C^1-function θ, then

$$\alpha'(s) \times \alpha''(s) = (0, 0, \theta'(s)),$$

so the z-component of $\alpha' \times \alpha''$ quantifies the rate of change of the angle describing the direction of the tangent vector.

The following lemma will allow us to introduce a measure for this 'bending' that does not depend on having a unit speed parametrisation.

Lemma 8.47 *Let $t \mapsto \alpha(t) = (x(t), y(t))$ be a regular C^2-curve. The quantity*

$$\boxed{\kappa(t) := \frac{\dot{x}(t)\ddot{y}(t) - \dot{y}(t)\ddot{x}(t)}{(\dot{x}^2(t) + \dot{y}^2(t))^{3/2}}}$$

is invariant under orientation-preserving C^2-reparametrisations of the curve.

Proof If $\phi\colon\ s \mapsto t(s)$ is a reparametrisation, and $\tilde{\alpha} := \alpha \circ \phi$, then

$$\begin{aligned}
\tilde{\alpha}' &= (\dot{\alpha} \circ \phi) \cdot \phi', \\
\tilde{\alpha}'' &= (\ddot{\alpha} \circ \phi) \cdot (\phi')^2 + (\dot{\alpha} \circ \phi) \cdot \phi''.
\end{aligned}$$

So the numerator of κ transforms as follows:

$$\begin{aligned}
\tilde{x}'\tilde{y}'' - \tilde{y}'\tilde{x}'' &= (\dot{x} \circ \phi) \cdot \phi' \cdot ((\ddot{y} \circ \phi) \cdot (\phi')^2 + (\dot{y} \circ \phi) \cdot \phi'') \\
&\quad - (\dot{y} \circ \phi) \cdot \phi' \cdot ((\ddot{x} \circ \phi) \cdot (\phi')^2 + (\dot{x} \circ \phi) \cdot \phi'') \\
&= ((\dot{x} \circ \phi)(\ddot{y} \circ \phi) - (\dot{y} \circ \phi)(\ddot{x} \circ \phi)) \cdot (\phi')^3.
\end{aligned}$$

For $\phi' > 0$, the denominator of κ likewise picks up a factor $(\phi')^3$. Hence $\tilde{\kappa} = \kappa \circ \phi$. □

Observe that, for a unit speed parametrisation $s \mapsto \alpha(s)$, the value $\kappa(s)$ equals the z-component of $\alpha'(s) \times \alpha''(s)$. Thus, our discussion motivates the following definition.

Definition 8.48 For a regular C^2-curve $t \mapsto \alpha(t) = (x(t), y(t))$ in the plane, the quantity $\kappa(t)$ as defined in Lemma 8.47 is called the **curvature** of α at the point $\alpha(t)$.

As we have seen, circles have constant curvature. In fact, this characterises circles.

Lemma 8.49 *Let $t \mapsto \alpha(t) \in \mathbb{R}^2$ be a regular C^2-curve of constant curvature $\kappa \neq 0$. Then the trace of α lies in a circle of radius $1/|\kappa|$.*

Proof It is convenient to identify $\mathbb{R}^2$ with $\mathbb{C}$, so that rotation through an angle $\pi/2$ can simply be written as multiplication by i. Without loss of generality we may assume that α has been parametrised as a unit speed curve. Then the acceleration α'' is orthogonal to the unit vector α', and it points to the left of α' precisely if $\kappa > 0$. Hence α'' can be written as $\alpha'' = \mathrm{i}\kappa\alpha'$. Set

$$\mathbf{p} := \alpha + \frac{\mathrm{i}}{\kappa}\alpha'.$$

Then $\mathbf{p}' = \alpha' + \mathrm{i}\alpha''/\kappa = \mathbf{0}$. Since $|\alpha - \mathbf{p}| = |\mathrm{i}\alpha'/\kappa| = 1/|\kappa|$, the trace of α lies on a circle of radius $1/|\kappa|$ about the centre $\mathbf{p}$. □

8.6.3 The duality between conics and circles

The following proposition shows the relevance of curvature for the discussion of polar reciprocals.

Proposition 8.50 *Let α be a C^2-curve in $\mathbb{R}^2$ with nowhere vanishing angular momentum, so that the polar reciprocal α^* is defined. If the curvature of α vanishes nowhere, then the polar reciprocal of α^* is also defined and equals α; in short $(\alpha^*)^* = \alpha$.*

Proof Write $\alpha(t) = (x(t), y(t))$ as before. By Proposition 8.46 we have

$$\alpha^*(t) = \frac{1}{c(t)}\,(\dot{y}(t), -\dot{x}(t)) =: (u(t), v(t)).$$

We compute

$$\dot{u} = \frac{\ddot{y}(x\dot{y} - y\dot{x}) - \dot{y}(x\ddot{y} - y\ddot{x})}{(x\dot{y} - y\dot{x})^2}$$
$$= -y\,\frac{\dot{x}\ddot{y} - \dot{y}\ddot{x}}{c^2}$$

and

$$\dot{v} = x\,\frac{\dot{x}\ddot{y} - \dot{y}\ddot{x}}{c^2},$$

whence

$$c^* := u\dot{v} - v\dot{u} = \frac{(x\dot{y} - y\dot{x})(\dot{x}\ddot{y} - \dot{y}\ddot{x})}{c^3} = \frac{\dot{x}\ddot{y} - \dot{y}\ddot{x}}{c^2}.$$

So the condition for $\boldsymbol{\alpha}^*$ to have non-zero angular momentum is precisely the condition for $\boldsymbol{\alpha}$ to have non-zero curvature. One then computes the polar reciprocal of $\boldsymbol{\alpha}^*$ as $(\dot{v}, -\dot{u})/c^* = (x, y)$. □

The duality expressed in this proposition is really a direct consequence of Lemma 8.44, see Exercise 8.22.

Proposition 8.51 *The polar reciprocal of a conic section $\mathcal{S} \subset \mathbb{R}^2$ with one focus at the origin lies on a circle. More precisely, the following statements hold.*

(e) *If $\mathcal{S}$ is an ellipse, the polar reciprocal is a circle containing the origin O in its interior.*
(p) *If $\mathcal{S}$ is a parabola, the polar reciprocal equals $C \setminus \{O\}$, where C is a circle passing through O.*
(h) *If $\mathcal{S}$ is a hyperbola, the polar reciprocal lies on a circle C with O in its exterior. Write A_1, A_2 for the points of contact of the two tangents from O to C. Then the two branches of the hyperbola correspond to the two arcs of $C \setminus \{A_1, A_2\}$.*

Notice that the points we had to remove from the circle C are those where the angular momentum with respect to O of any regular parametrisation of C is zero.

Both from the geometric proof below and from the analytic proof in Exercise 8.26 it should become clear that any circle in the plane can be realised as the polar reciprocal of a suitable conic with one focus at the origin. Then, with Proposition 8.50, it follows that the polar reciprocal of any circle (with one or two points removed, as the case may be) will be such a conic. This is the key to proving Kepler's first theorem.

Proof of Proposition 8.51 We give a geometric proof of case (e). Geometric proofs of the other two cases and an analytic proof covering all three cases will be discussed in the exercises.

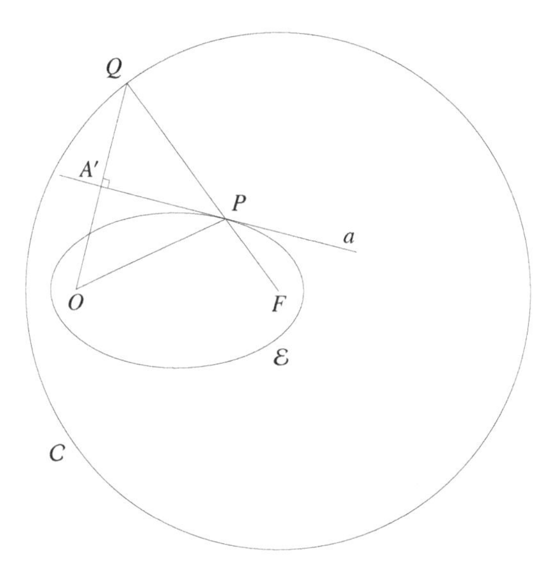

Figure 8.15 The duality between ellipses and circles.

Let $\mathcal{E}$ be an ellipse with foci at O and F; possibly $F = O$. Let P be a point on $\mathcal{E}$, and let a be the tangent line to $\mathcal{E}$ at P. Let Q be the mirror image of O in a. By the reflection property of ellipses, see Exercise 3.9, the points F, P, Q lie on a straight line. By construction, OQ intersects a orthogonally at a point A'. See Figure 8.15.

As P moves along $\mathcal{E}$, the point Q describes a circle about F of radius equal to the major axis of $\mathcal{E}$. The point A' is obtained from Q by the homothety $(x, y) \mapsto (x/2, y/2)$ of $\mathbb{R}^2$. Consequently, A' moves on a circle about the midpoint of O and F (the centre of $\mathcal{E}$) of radius equal to the semi-major axis. Notice that this circle contains O in its interior. The pole A of a is obtained by inversion of A'. By Proposition 8.10, A likewise describes a circle. □

8.6.4 Proof of Kepler's first law

Let $\mathbf{r}\colon t \mapsto (x(t), y(t))$ be a solution of the Kepler problem, i.e.

$$\ddot{x} = -\frac{\mu x}{r^3}, \qquad \ddot{y} = -\frac{\mu y}{r^3}.$$

By Proposition 1.3, the angular momentum of $\mathbf{r}$ is constant:

$$x\dot{y} - y\dot{x} = c.$$

For $c \neq 0$, the polar reciprocal $\mathbf{r}^*$ is given by

$$t \longmapsto \frac{1}{c}\,(\dot{y}(t), -\dot{x}(t)) =: (X(t), Y(t)).$$

Then

$$\left.\begin{aligned} \dot{X} &= \frac{\ddot{y}}{c} = -\frac{\mu}{c}\cdot\frac{y}{r^3} \\ \dot{Y} &= -\frac{\ddot{x}}{c} = \frac{\mu}{c}\cdot\frac{x}{r^3} \end{aligned}\right\}$$

and

$$\left.\begin{aligned} \ddot{X} &= \frac{\mu}{c}\cdot\frac{-r\dot{y} + 3y\dot{r}}{r^4} \\ \ddot{Y} &= \frac{\mu}{c}\cdot\frac{r\dot{x} - 3x\dot{r}}{r^4}. \end{aligned}\right\}$$

Notice that, *a priori*, the polar reciprocal of a C^k-curve is only of differentiability class C^{k-1}. A C^2-solution of the Kepler problem (K), however, is automatically C^∞, since (K) implies that $\ddot{\mathbf{r}}$ has the same class of differentiability as $\mathbf{r}$. So $\mathbf{r}^*$ will be likewise of class C^∞.

The curvature of $\mathbf{r}^*$ is constant:

$$\frac{\dot{X}\ddot{Y} - \dot{Y}\ddot{X}}{(\dot{X}^2 + \dot{Y}^2)^{3/2}} = \frac{c}{\mu}\,(x\dot{y} - y\dot{x}) = \frac{c^2}{\mu}.$$

By Lemma 8.49, the curve $\mathbf{r}^*$ lies on a circle. Then Proposition 8.51 (with the comments following it) gives us Kepler's first law.

Remark 8.52 The fact that the polar reciprocal $\mathbf{r}^*$ of a solution $\mathbf{r}$ of the Kepler problem lies on a circle can also be seen by the argument in our proof of Proposition 8.3. We showed there by integrating the Kepler equation that the velocity vector $\dot{\mathbf{r}} = (\dot{x}, \dot{y})$ lies on a circle. Hence so does the polar reciprocal $\mathbf{r}^* = (\dot{y}, -\dot{x})/c$, since c is constant.

8.6.5 The projective viewpoint

In the construction of the polar reciprocal of a curve α we had to exclude points on α with vanishing angular momentum. The reason is that Lemma 8.44 holds only for points $A \neq O$ and lines not passing through O. In this section I want to show that by enlarging $\mathbb{R}^2$ to the so-called projective plane $\mathbb{RP}^2$ one can give a more natural and unified definition of the dual curve α^*.

As a set, the projective plane $\mathbb{RP}^2$ is defined to be the set of radial lines in $\mathbb{R}^3$. In order to give a topology to this space, one defines it more formally as a quotient space. There are two ways to go about this. One way is to observe that any radial line in $\mathbb{R}^3$ intersects the unit 2-sphere S^2 at a pair of antipodal points; conversely, any such pair determines a radial line. Alternatively, we notice that any point in $\mathbb{R}^3 \setminus \{\mathbf{0}\}$ determines a radial line, and two points determine the same radial line if and only if they differ by multiplication with a scalar $\lambda \in \mathbb{R}^\times := \mathbb{R} \setminus \{0\}$.

Definition 8.53 The **projective plane** $\mathbb{RP}^2$ is the quotient space

$$\mathbb{R}^3 \setminus \{\mathbf{0}\}/\sim,$$

where the equivalence relation $\sim$ is defined by

$$\mathbf{x} \sim \mathbf{y} \quad :\Longleftrightarrow \quad \text{There is a } \lambda \in \mathbb{R}^\times \text{ such that } \mathbf{y} = \lambda \mathbf{x}.$$

Write $[\mathbf{x}] \in \mathbb{RP}^2$ for the equivalence class of $\mathbf{x} \in \mathbb{R}^3 \setminus \{\mathbf{0}\}$. We then have the projection mapping

$$\begin{array}{rccc} \pi: & \mathbb{R}^3 \setminus \{\mathbf{0}\} & \longrightarrow & \mathbb{RP}^2 \\ & \mathbf{x} & \longmapsto & [\mathbf{x}]. \end{array}$$

The **quotient topology** on $\mathbb{RP}^2$ is defined by

$$U \subset \mathbb{RP}^2 \text{ is open} \quad :\Longleftrightarrow \quad \pi^{-1}(U) \subset \mathbb{R}^3 \setminus \{\mathbf{0}\} \text{ is open.}$$

This does indeed satisfy the axioms of a topology, see Exercise 8.30, and it is the finest topology on $\mathbb{RP}^2$ for which the quotient map π is continuous.

Remark 8.54 One arrives at the same topology on $\mathbb{RP}^2$ if one equips $S^2 \subset \mathbb{R}^3$ with the induced topology (see Exercise 8.3) and then

$$\mathbb{RP}^2 = S^2/\mathbf{x} \sim -\mathbf{x}$$

with the quotient topology. This has the advantage, for instance, that one immediately recognises $\mathbb{RP}^2$ as a compact topological space, see the reasoning in the proof of Theorem 10.9. Moreover, the quotient map $S^2 \to \mathbb{RP}^2$ is simply a double covering, which is a more convenient setting for discussing curves in $\mathbb{RP}^2$.

Now label the cartesian coordinates of $\mathbb{R}^3$ by x_0, x_1, x_2. Any non-horizontal radial line in $\mathbb{R}^3$, that is, one not contained in the x_1x_2-plane, intersects the affine plane $\{x_0 = 1\}$ in a unique point, see Figure 8.16. This allows us to identify $\mathbb{R}^2$ as a subset of $\mathbb{RP}^2$. So we may think of $\mathbb{RP}^2$ as a copy of $\mathbb{R}^2$ together with one additional point for each radial line in $\mathbb{R}^2$, *viz.*, the horizontal radial lines in $\mathbb{R}^3$. These additional points are sometimes referred to as the **points at infinity**.

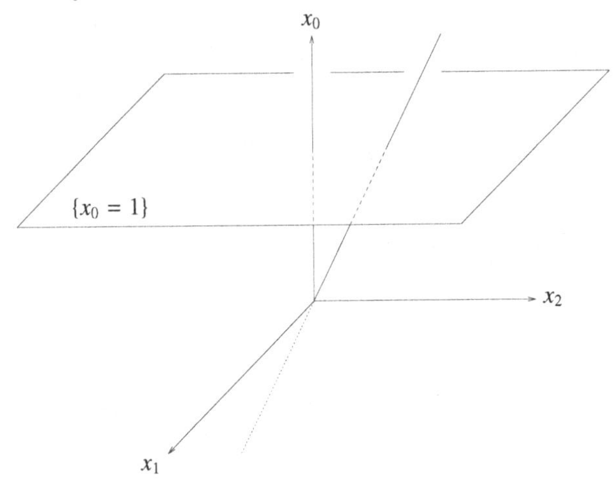

Figure 8.16 $\mathbb{R}^2$ as a subset of $\mathbb{RP}^2$.

Completely analogously, for any $n \in \mathbb{N}$ one defines the **projective space** $\mathbb{RP}^n$ of dimension n as the space of radial lines in $\mathbb{R}^{n+1}$. In particular, $\mathbb{RP}^1$ consists of a copy of $\mathbb{R}$ together with a single point at infinity, and the points at infinity in $\mathbb{RP}^2$ constitute a copy of $\mathbb{RP}^1$.

Points in $\mathbb{RP}^2$ can be described in terms of so-called **homogeneous coordinates**. If $\mathbf{x} = (x_0, x_1, x_2) \in \mathbb{R}^3 \setminus \{\mathbf{0}\}$ represents a point $[\mathbf{x}] \in \mathbb{RP}^2$, we write that point in coordinates as $[x_0, x_1, x_2]$. The name 'homogeneous' coordinates stems from the fact that

$$[x_0, x_1, x_2] = [\lambda x_0, \lambda x_1, \lambda x_2] \quad \text{for } \lambda \in \mathbb{R}^\times.$$

In these coordinates, the inclusion of $\mathbb{R}^2$ as a subset of $\mathbb{RP}^2$, as defined geometrically in Figure 8.16, corresponds to the map

$$\mathbb{R}^2 \ni (\xi_1, \xi_2) \longmapsto [1, \xi_1, \xi_2] \in \mathbb{RP}^2.$$

A **(projective) line** in $\mathbb{RP}^2$ is defined as a radial plane in $\mathbb{R}^3$ or, more precisely, as the set of points in $\mathbb{RP}^2$ corresponding to radial lines in $\mathbb{R}^3$ inside a single given radial plane. The radial plane in $\mathbb{R}^3$ orthogonal to a non-zero vector (a_0, a_1, a_2) is given by the equation

$$a_0 x_0 + a_1 x_1 + a_2 x_2 = 0.$$

This equation stays unchanged if (a_0, a_1, a_2) is replaced by $\lambda(a_0, a_1, a_2)$ for some $\lambda \in \mathbb{R}^\times$. So the projective lines in $\mathbb{RP}^2$ form another copy of $\mathbb{RP}^2$. We call this the **dual projective plane** $(\mathbb{RP}^2)^*$ and write points in it in homogeneous coordinates as $[a_0, a_1, a_2]^*$.

We regard $\mathbb{R}^2$ as a subset of $(\mathbb{RP}^2)^*$ via the inclusion

$$(\alpha_1, \alpha_2) \longmapsto [-1, \alpha_1, \alpha_2]^*,$$

where the choice of sign will be justified by the discussion that follows.

Given a set of points $[x_0, x_1, x_2]$ in $\mathbb{RP}^2$, we can view the points with $x_0 \neq 0$ as points in $\mathbb{R}^2$ via $(\xi_1, \xi_2) := (x_1/x_0, x_2/x_0)$. These points constitute the **affine part** of the given subset of $\mathbb{RP}^2$. In $(\mathbb{RP}^2)^*$ we define the affine part via $(\alpha_1, \alpha_2) := -(a_1/a_0, a_2/a_0)$.

We now want to extend the notions of pole and polar to this projective setting. As we shall see, Lemma 8.44 then becomes virtually tautological. Any line in $\mathbb{R}^2$ can be written in the form

$$a = \{(\xi_1, \xi_2) \in \mathbb{R}^2 \colon \ a_0 + a_1\xi_1 + a_2\xi_2 = 0\}$$

with $a_1^2 + a_2^2 = 1$. Notice that the unit vector (a_1, a_2) is orthogonal to the line a. So the point

$$A' := -a_0\,(a_1, a_2) \in a$$

is the foot of the perpendicular from O to a. The condition for a not to contain the origin O is $a_0 \neq 0$; then

$$A := -\frac{1}{a_0}\,(a_1, a_2)$$

is the pole of a.

The projective extension of a, i.e. the projective line whose affine part is a, is given by homogenising the equation defining a:

$$\{(x_0, x_1, x_2) \in \mathbb{R}^3 \colon \ a_0x_0 + a_1x_1 + a_2x_2 = 0\}.$$

We continue to write a for this projective line.

Under the inclusion $\mathbb{R}^2 \subset (\mathbb{RP}^2)^*$, the point $A \in \mathbb{R}^2$ maps to

$$[-1, -a_1/a_0, -a_2/a_0]^* = [a_0, a_1, a_2]^*.$$

Again, we continue to write A for this point; this point $A \in (\mathbb{RP}^2)^*$ is now also defined for $a_0 = 0$. Observe that $A \in (\mathbb{RP}^2)^*$ is simply the point defining the line $a \subset \mathbb{RP}^2$.

Dually, we can start from a line $b \subset \mathbb{R}^2 \subset (\mathbb{RP}^2)^*$, which after extension to a line in $(\mathbb{RP}^2)^*$ may be regarded as a point $B \in \mathbb{RP}^2$. The same reasoning as above shows that B is actually the pole of b under the inclusion $\mathbb{R}^2 \subset \mathbb{RP}^2$, provided b did not contain O, see Exercise 8.32.

This discussion shows that the projective extension of Lemma 8.44 is the following statement.

Lemma 8.55 *Let* $A = [a_0, a_1, a_2]^*$ *be a point in* $(\mathbb{R}\mathrm{P}^2)^*$*, defining the line* $a \subset \mathbb{R}\mathrm{P}^2$*, and let* $B = [b_0, b_1, b_2]$ *be a point in* $\mathbb{R}\mathrm{P}^2$*, defining the line* $b \subset (\mathbb{R}\mathrm{P}^2)^*$*. Then* $A \in b$ *if and only if* $B \in a$*.*

Proof Both statements $A \in b$ and $B \in a$ are equivalent to the equation

$$a_0 b_0 + a_1 b_1 + a_2 b_2 = 0$$

being satisfied. □

Next we discuss curves in the projective plane. A curve α in $\mathbb{R}\mathrm{P}^2$ can be written in homogeneous coordinates as

$$t \longmapsto \alpha(t) = [\mathbf{x}(t)] = [x_0(t), x_1(t), x_2(t)], \quad t \in I.$$

We say that α is of class C^k if we find a curve $\mathbf{x}$ in $\mathbb{R}^3 \setminus \{\mathbf{0}\}$ of class C^k representing $\alpha = [\mathbf{x}]$.

Definition 8.56 A C^1-curve $t \mapsto \alpha(t) = [\mathbf{x}(t)]$ in $\mathbb{R}\mathrm{P}^2$ is **regular** if the two vectors $\mathbf{x}(t), \dot{\mathbf{x}}(t) \in \mathbb{R}^3$ are linearly independent for all $t \in I$.

Observe that, if $\mathbf{x}$ is replaced by $\lambda\mathbf{x}$ for some C^1-function $\lambda \colon I \to \mathbb{R}^\times$, then the second vector in question changes to $\dot{\lambda}\mathbf{x} + \lambda\dot{\mathbf{x}}$, so the definition of regularity does not depend on the specific choice of a C^1-representative $\mathbf{x}$ of α. By choosing $\lambda = |\mathbf{x}|^{-1}$, we may always assume that $|\mathbf{x}| = 1$, i.e. $\mathbf{x}$ describes a curve on the unit sphere $S^2 \subset \mathbb{R}^3$. By differentiating the equation $\langle \mathbf{x}, \mathbf{x} \rangle = 1$ we obtain $\langle \mathbf{x}, \dot{\mathbf{x}} \rangle = 0$, i.e. $\dot{\mathbf{x}}$ is orthogonal to $\mathbf{x}$ (and hence tangent to S^2, as one should expect). For this specific parametrisation, regularity simply means $\dot{\mathbf{x}} \neq \mathbf{0}$, i.e. $\mathbf{x}$ is a regular curve on S^2.

The following definition is then likewise independent of the specific choice of representative $\mathbf{x}$.

Definition 8.57 The **tangent line** of a regular C^1-curve $\alpha = [\mathbf{x}]$ in $\mathbb{R}\mathrm{P}^2$ at the point $\alpha(t)$ is the projective line determined by the radial plane in $\mathbb{R}^3$ spanned by $\mathbf{x}(t)$ and $\dot{\mathbf{x}}(t)$.

Thanks to the cross product on $\mathbb{R}^3$, the tangent line of a regular curve $\alpha = [\mathbf{x}]$ permits a simple explicit description, since the plane spanned by $\mathbf{x}(t)$ and $\dot{\mathbf{x}}(t)$ is the same as the radial plane orthogonal to $\mathbf{x}(t) \times \dot{\mathbf{x}}(t)$. Notice that $(\lambda\mathbf{x}) \times \frac{\mathrm{d}}{\mathrm{d}t}(\lambda\mathbf{x}) = \lambda^2 \mathbf{x} \times \dot{\mathbf{x}}$.

Definition 8.58 The **dual curve** of a regular curve $\alpha = [\mathbf{x}]$ in $\mathbb{R}\mathrm{P}^2$ is the curve $\alpha^* := [\mathbf{x} \times \dot{\mathbf{x}}]^*$ in $(\mathbb{R}\mathrm{P}^2)^*$.

We now want to show how the dual curve relates to the polar reciprocal. Let $t \mapsto \alpha(t) = [\mathbf{x}(t)] = [x_0(t), x_1(t), x_2(t)]$ be a regular curve in $\mathbb{R}\mathrm{P}^2$. Then

$$\alpha^* = [\mathbf{x} \times \dot{\mathbf{x}}]^* = [x_1\dot{x}_2 - x_2\dot{x}_1, x_2\dot{x}_0 - x_0\dot{x}_2, x_0\dot{x}_1 - x_1\dot{x}_0]^*.$$

For points $[1, x, y]$ in the affine part of α this becomes

$$[x\dot{y} - y\dot{x}, -\dot{y}, \dot{x}]^*,$$

and if the angular momentum $c = x\dot{y} - y\dot{x}$ at a point in the affine part of α does not vanish, the corresponding point of α^* lies in the affine part and can be written as

$$[-1, \dot{y}/c, -\dot{x}/c]^*.$$

So these points in the affine part of α^* constitute the polar reciprocal.

The projective analogue of Proposition 8.50 is the following.

Proposition 8.59 *Let $\alpha = [\mathbf{x}]$ be a C^2-curve in $\mathbb{R}\mathrm{P}^2$ with $\mathbf{x}, \dot{\mathbf{x}}, \ddot{\mathbf{x}}$ linearly independent in $\mathbb{R}^3$ at every point of the curve. Then the dual of α^* is defined and equals α.*

Proof By definition we have $\alpha^* = [\mathbf{x} \times \dot{\mathbf{x}}]^*$. The condition for this dual curve to be regular is that $\mathbf{x} \times \dot{\mathbf{x}}$ and $\frac{\mathrm{d}}{\mathrm{d}t}(\mathbf{x} \times \dot{\mathbf{x}}) = \mathbf{x} \times \ddot{\mathbf{x}}$ be linearly independent. We compute with the Graßmann identity (1.1) from Chapter 1:

$$(\mathbf{x} \times \dot{\mathbf{x}}) \times (\mathbf{x} \times \ddot{\mathbf{x}}) = \langle \mathbf{x}, \mathbf{x} \times \ddot{\mathbf{x}} \rangle \dot{\mathbf{x}} - \langle \dot{\mathbf{x}}, \mathbf{x} \times \ddot{\mathbf{x}} \rangle \mathbf{x} = -\langle \dot{\mathbf{x}}, \mathbf{x} \times \ddot{\mathbf{x}} \rangle \mathbf{x}.$$

The triple product $-\langle \dot{\mathbf{x}}, \mathbf{x} \times \ddot{\mathbf{x}} \rangle$ equals the determinant $\det(\mathbf{x}, \dot{\mathbf{x}}, \ddot{\mathbf{x}})$ or, geometrically, the oriented volume of the parallelepiped spanned by $\mathbf{x}, \dot{\mathbf{x}}$ and $\ddot{\mathbf{x}}$.

So the regularity condition for α^* is equivalent to the non-vanishing of that determinant. Then the previous computation shows that the dual of α^* is represented by a non-vanishing multiple of $\mathbf{x}$, and hence equals α. □

If $[\mathbf{x}]$ is a regular curve in $\mathbb{R}\mathrm{P}^2$, we may choose $\mathbf{x}$ to be a unit speed curve $s \mapsto \mathbf{x}(s)$, $s \in I$, in the 2-sphere S^2. Then the acceleration $\mathbf{x}''$ is orthogonal to the unit velocity vector $\mathbf{x}'$. Write $\mathbf{n}$ for the outer normal unit vector to S^2, which happens to coincide with $\mathbf{x}$. The so-called **intrinsic normal vector $\mathbf{S} := \mathbf{n} \times \mathbf{x}'$** is a vector field along the curve, tangent to S^2 but orthogonal to the curve. The acceleration $\mathbf{x}''$ can then be written as

$$\mathbf{x}'' = \kappa_{\mathrm{g}}\mathbf{S} + \kappa_{\mathrm{n}}\mathbf{n}.$$

Here κ_{g} and κ_{n} are real-valued functions on I, called the **geodesic curvature** and **normal curvature**, respectively.

The condition on $\mathbf{x}, \mathbf{x}', \mathbf{x}''$ to be linearly independent is then seen to be equivalent to the geodesic curvature being nowhere zero.[14]

8.7 Newton's vs. Hooke's law

In this section we study solutions of the Kepler problem in a fixed plane, which we identify with the complex plane $\mathbb{C}$. I wish to present a duality between, on the one hand, the Kepler solutions and, on the other, solutions of the much simpler linear equation

$$\mathbf{w}'' = -\lambda \mathbf{w}. \tag{H}$$

We are looking for C^2-maps $s \mapsto \mathbf{w}(s) \in \mathbb{C}$ that solve equation (H), where λ is a real parameter, and the prime denotes the derivative with respect to the variable s.[15] Equation (H) with a parameter $\lambda > 0$ describes Hooke's law for springs: the force required to extend or compress a spring by a certain distance is proportional to that distance.

The solutions of (H), which are always defined on the full real line, are as follows. For $\lambda = 0$, the solutions are linear: $s \mapsto \mathbf{a}s + \mathbf{b}$, with $\mathbf{a}, \mathbf{b} \in \mathbb{C}$. By a rotation of the complex plane and a shift in the s-parameter it suffices to consider solutions of the form

$$\mathbf{w}(s) = as + ib \tag{H$_0$}$$

with $a, b \in \mathbb{R}$.

For $\lambda > 0$, a fundamental system of solutions is given by $\cos(\sqrt{\lambda}\, s)$ and $\sin(\sqrt{\lambda}\, s)$, i.e. the general solution is a complex linear combination of these two. In particular, every solution is bounded, so it reaches a maximal distance a from the origin, where the velocity is orthogonal to the position vector. By a rotation of the complex plane and a shift in the s-parameter we may assume that $\mathbf{w}(0) = a$. Then $\mathbf{w}'(0) = ib\sqrt{\lambda}$ for some real parameter b, which we may assume to be non-negative by a reflection of $\mathbb{C}$, if necessary. The solution is then given by

$$\mathbf{w}(s) = a\cos(\sqrt{\lambda}\, s) + ib\sin(\sqrt{\lambda}\, s).$$

Our assumption that a be the maximal distance from the origin implies $b \leq a$, since $\mathbf{w}(\pi/2\sqrt{\lambda}) = ib$. We see that, for $b > 0$, the solution curve is an ellipse

[14] A geodesic on a surface would be a curve on that surface whose geodesic curvature vanishes identically, see Exercises 8.9 and 8.10.

[15] In contrast with the preceding section, here s does not, in general, correspond to arc length parametrisation.

centred at the origin, with semi-major axis a along the real axis and semi-minor axis b along the imaginary axis. This solution can be rewritten as

$$\mathbf{w}(s) = p \exp(\mathrm{i}\sqrt{\lambda}\, s) + q \exp(-\mathrm{i}\sqrt{\lambda}\, s) \tag{H$_+$}$$

with $p = (a+b)/2$ and $q = (a-b)/2$, i.e. $p > q \geq 0$. Notice that the foci are at the points $(\pm\sqrt{a^2-b^2}, 0) = (\pm 2\sqrt{pq}, 0)$.

Similar considerations, see Exercise 8.33, show that for $\lambda < 0$ the solution can be written, without loss of generality, as

$$\mathbf{w}(s) = a \cosh(\sqrt{-\lambda}\, s) + \mathrm{i} b \sinh(\sqrt{-\lambda}\, s)$$

with $a, b \in \mathbb{R}_0^+$, or

$$\mathbf{w}(s) = \gamma \exp(\sqrt{-\lambda}\, s) + \overline{\gamma} \exp(-\sqrt{-\lambda}\, s), \tag{H$_-$}$$

where $\gamma = (a + \mathrm{i}b)/2$. For $a, b > 0$ this is a parametrisation of the hyperbola

$$\left\{(x, y) \in \mathbb{R}^2 : \frac{x^2}{a^2} - \frac{y^2}{b^2} = 1\right\}$$

centred at the origin and with foci at the points $(\pm\sqrt{a^2+b^2}, 0) = (\pm 2|\gamma|, 0)$, see Exercise 2.5.

Hooke's equation is a special case of the central force problem, with a conservative force field given by the potential $\lambda|\mathbf{w}|^2/2$. Thus, as in Chapter 1 we have two preserved quantities, the angular momentum $c_{\mathrm{H}} = |\mathbf{w} \times \mathbf{w}'|$ and the energy

$$h_{\mathrm{H}} = \frac{1}{2}(|\mathbf{w}'|^2 + \lambda|\mathbf{w}|^2).$$

The following proposition relates solutions of the Hooke problem (H) to those of the Kepler problem

$$\ddot{\mathbf{z}} = -\frac{\mu}{|\mathbf{z}|^2} \cdot \frac{\mathbf{z}}{|\mathbf{z}|} \tag{K}$$

in $\mathbb{C} \setminus \{\mathbf{0}\}$.

Proposition 8.60 *Let $s \mapsto \mathbf{w}(s) \in \mathbb{C} \setminus \{\mathbf{0}\}$ be a solution of* (H) *with $c_{\mathrm{H}} \neq 0$. Define the parameter $t = t(s)$ by*

$$t(s) := \int_0^s |\mathbf{w}(\sigma)|^2 \,\mathrm{d}\sigma.$$

Because of $t'(s) = |\mathbf{w}(s)|^2 \neq 0$ we may then, conversely, regard s as a function of t. Set

$$\mathbf{z}(t) := \mathbf{w}^2(s(t)).$$

Then $\mathbf{z}$ *is a solution of* (K) *with* $\mu = 4h_{\mathrm{H}}$. *The angular momentum and the energy of this solution are given by* $c_{\mathrm{K}} = 2c_{\mathrm{H}}$ *and* $h_{\mathrm{K}} = -2\lambda$.

Proof For a differentiable function $t \mapsto f(t) = g(s(t))$ we have

$$\frac{\mathrm{d}f}{\mathrm{d}t} = \frac{\mathrm{d}g}{\mathrm{d}s} \cdot \frac{\mathrm{d}s}{\mathrm{d}t} = \frac{1}{|\mathbf{w}(s(t))|^2} \cdot \frac{\mathrm{d}g}{\mathrm{d}s}.$$

We write this more briefly as

$$\frac{\mathrm{d}f}{\mathrm{d}t} = \frac{1}{|\mathbf{w}|^2} \cdot \frac{\mathrm{d}g}{\mathrm{d}s}.$$

Applied to $\mathbf{z}(t) = \mathbf{w}^2(s(t))$, this yields

$$\dot{\mathbf{z}} = \frac{1}{|\mathbf{w}|^2} \cdot \frac{\mathrm{d}\mathbf{w}^2}{\mathrm{d}s} = \frac{2\mathbf{w}'}{\overline{\mathbf{w}}}$$

and

$$\begin{aligned}
\ddot{\mathbf{z}} &= \frac{1}{|\mathbf{w}|^2} \cdot \frac{\mathrm{d}}{\mathrm{d}s}\left(\frac{2\mathbf{w}'}{\overline{\mathbf{w}}}\right) \\
&= \frac{2}{|\mathbf{w}|^2} \cdot \left(-\frac{\overline{\mathbf{w}}'\mathbf{w}'}{\overline{\mathbf{w}}^2} + \frac{\mathbf{w}''}{\overline{\mathbf{w}}}\right) \\
&= \frac{-2}{\mathbf{w}\overline{\mathbf{w}}^3} \cdot (|\mathbf{w}'|^2 + \lambda|\mathbf{w}|^2) \\
&= -\frac{\mu}{\mathbf{w}\overline{\mathbf{w}}^3} = -\mu\frac{\mathbf{z}}{|\mathbf{z}|^3},
\end{aligned}$$

where we have set $\mu = 4h_{\mathrm{H}}$. For the angular momentum we compute

$$|\mathbf{z} \times \dot{\mathbf{z}}| = \left|\mathbf{w}^2 \times \frac{2\mathbf{w}\mathbf{w}'}{|\mathbf{w}|^2}\right| = 2|\mathbf{w} \times \mathbf{w}'|,$$

where we have used that multiplication by $\mathbf{w}$ is a conformal map $\mathbb{C} \to \mathbb{C}$. Observe that our choice of reparametrisation $t = t(s)$ is, up to an arbitrary constant, precisely what makes the transformed curve $t \mapsto \mathbf{z}(t)$ again of constant angular momentum.

Finally, the energy is given by

$$\begin{aligned}
h_{\mathrm{K}} &= \frac{1}{2}|\dot{\mathbf{z}}|^2 - \frac{\mu}{|\mathbf{z}|} \\
&= 2\frac{|\mathbf{w}'|^2}{|\mathbf{w}|^2} - \frac{2(|\mathbf{w}'|^2 + \lambda|\mathbf{w}|^2)}{|\mathbf{w}|^2} \\
&= -2\lambda.
\end{aligned}$$

□

Remark 8.61 The argument in the preceding proof works just as well in the case $c_{\mathrm{H}} = 0$; a problem arises only for those values of s where $\mathbf{w}(s) = 0$, since here the function $s \mapsto t(s)$ is not invertible, and the speed of $t \mapsto \mathbf{z}(t)$ diverges to infinity.

Up to inessential choices, the solutions of (H) with $c_{\mathrm{H}} = 0$ are given by

$$s \mapsto \begin{cases} as & \text{for } \lambda = 0, \\ a\cos(\sqrt{\lambda}\, s) & \text{for } \lambda > 0, \\ a\cosh(\sqrt{-\lambda}\, s) + b\sinh(\sqrt{-\lambda}\, s) & \text{for } \lambda < 0, \end{cases}$$

where $a, b \in \mathbb{R}_0^+$. These are simply parametrisations of (parts of) the real line. When we apply the map $\mathbf{w} \mapsto \mathbf{w}^2$ to these solutions, we obtain (in the case $\mu = 4h_{\mathrm{H}} > 0$) the regularised bounce orbits we discussed in Sections 8.3 and 8.5, see Exercise 8.34.

So far we have shown only that solutions of (H) give rise to solutions of (K). In fact, all solutions of (K) arise in this manner.

Proposition 8.62 *Given any solution* $\mathbf{z}$ *of* (K)*, there is a solution* $\mathbf{w}$ *of* (H) *such that* $\mathbf{z}(t) = \mathbf{w}^2(s(t))$*, with* s *and* t *related as in Proposition 8.60.*

Proof From a solution $\mathbf{w}$ of (H), the parameter μ and the initial values of the corresponding solution $\mathbf{z}$ of (K) can be computed by

$$(\mathbf{z}(0), \dot{\mathbf{z}}(0), \mu) = \left(\mathbf{w}^2(0), \frac{2}{\overline{\mathbf{w}(0)}} \cdot \mathbf{w}'(0), 2(|\mathbf{w}'(0)|^2 + \lambda|\mathbf{w}(0)|^2)\right). \tag{8.13}$$

A solution $\mathbf{z}$ of the Kepler problem (with given $\mu > 0$) is determined by its initial values $\mathbf{z}(0)$ and $\dot{\mathbf{z}}(0)$. Given any μ and choice of these initial values, choose $\mathbf{w}(0)$, $\mathbf{w}'(0)$ and λ that realise them in the sense of (8.13). The corresponding solution $\mathbf{w}$ of (H) then transforms to the given solution $\mathbf{z}$ of (K). □

It is now a simple matter to deduce, once again, Kepler's first law. By the relations in Proposition 8.60, a solution of (K) with $c_{\mathrm{K}} \neq 0$ and $h_{\mathrm{K}} < 0$ comes from a solution of (H) with $c_{\mathrm{H}} \neq 0$ and $\lambda > 0$, i.e. a solution of type (H_+). We obtain

$$\mathbf{z}(t(s)) = \mathbf{w}^2(s) = p^2 \exp(\mathrm{i} \cdot 2\sqrt{\lambda}\, s) + q^2 \exp(-\mathrm{i} \cdot 2\sqrt{\lambda}\, s) + 2pq,$$

which describes an ellipse with foci at $(\pm 2pq + 2pq, 0)$.

For the case $h_{\mathrm{K}} \geq 0$ see Exercise 8.35.

Notes and references

The hodograph was introduced and named by Hamilton (1847); the concept had been anticipated in §22 of (Möbius, 1843). Rather amazingly, Theorem 8.2 and the result discussed in Exercise 8.27 are proved in (Hamilton, 1847) without the use of a single mathematical symbol.

The concept of inversion was introduced by Steiner in a manuscript dated 'Sonntag, 8. Hornung [February] 1824' and reproduced in (Bützberger, 1913). Steiner termed it the theory of 'Wiedergeburt und Auferstehung' [rebirth and resurrection]. Strictly speaking, conformality of the inversion is not proved there, but Bützberger argues from circumstantial evidence that this must have been known to Steiner: "Nach der Veröffentlichung dieser Arbeit wird kein Zweifel mehr bestehen können, daß Steiner auch die Haupteigenschaft seiner 'Wiedergeburt und Auferstehung', die *Winkeltreue*, gekannt und benutzt hat."

Theorem 8.16 is due to Moser (1970). Theorem 8.33 was first announced by Osipov in 1972, in a paper published in Russian. It was proved again by Belbruno (1977), and Osipov took another look at the result in (Osipov, 1977). I largely followed the exposition in (Milnor, 1983), to which I have added details about the parametric versions of the results.

For a comprehensive and very readable introduction to hyperbolic geometry see (Anderson, 2005). Anderson's approach, like the one I followed, is analytic. This means that one works with a specific model of hyperbolic space, such as the half-space model with the hyperbolic metric we discussed. For a synthetic approach, where one starts from a set of axioms for hyperbolic geometry, see (Martin, 1975), which also contains a wealth of historical information about the discovery of this geometry.

The Levi-Civita parameter (Definition 8.40) was introduced by Levi-Civita (1920) for the purpose of regularising collisions. A more gentle account is (Levi-Civita, 1924). Regularisation of the Kepler problem is a subject of ongoing interest, see for instance (Heckman and de Laat, 2012) and (Hu and Santoprete, 2014).

The dual of Proposition 8.51 – the polar reciprocal of a circle is a conic with one focus at the origin – is proved synthetically in (Coxeter and Greitzer, 1967). Beware that they *define* conics as polar reciprocals of circles and then derive the description in terms of focus and directrix as a theorem.

The proof of Kepler's first law in Section 8.6.4 is due to Coolidge (1920). For more on the notions of geodesic and normal curvature of curves on surfaces in $\mathbb{R}^3$, the reader should consult a text on elementary differential geometry. Good examples are (Bär, 2010) and (Millman and Parker, 1977). The latter also contains material relevant for the discussion of the space forms in

Section 8.5, such as the definition of sectional curvature. The sectional curvature of the hyperbolic space forms is computed in (Millman and Parker, 1977, Exercise 8.13).

By Remark 8.52, Coolidge's reasoning is essentially equivalent to Hamilton's hodograph theorem, which he does not cite. Indeed, Hamilton proved Theorem 8.2 by showing that the hodograph has constant curvature, rather than integrating the Kepler equation as we have done. Hamilton's theorem has been rediscovered several times; see (Derbes, 2001), (van Haandel and Heckman, 2009) and (Ostermann and Wanner, 2012) for more historical information.

The duality result in Section 8.7 is implicit in (Newton, 1687) in 'Another solution' to Propositions 11 and 12 of Book 1, which deal with the inverse problem of deriving the force law $F \propto r^{-2}$ from the known elliptic or hyperbolic trajectory (with force centre at one of the foci); cf. the notes to Chapter 3. The first explicit statement of this duality is due to Bohlin (1911) and Kasner (1913). For a generalisation of this result see (Arnol'd and Vasil'ev, 1989; Needham, 1993; Hall and Josić, 2000) and the appendix in (Arnol'd, 1990), as well as Exercise 8.36. We shall return to this duality result in Section 9.1, see in particular Theorem 9.13, where we give a strikingly simple and very geometric proof, due to Arnol'd (1990), that any conformal transformation of $\mathbb{C}$ gives rise to dual force laws – here the attribute 'simple' presupposes knowledge of how to translate motions in a conservative force field into geodesics of a suitable metric, the so-called Jacobi metric. This is in fact the geometric interpretation of Maupertuis's least action principle, which we discuss in the next chapter.

An intriguing example is the force law $F \propto r^{-5}$, which is self-dual under the mapping $\mathbf{w} \mapsto \mathbf{w}^{-1}$, i.e. inversion in the unit circle followed by a reflection. Under this map, the straight lines not containing the origin are transformed into circles through the origin. The former may be regarded as solutions for the zero force field; the latter are then solutions for a non-zero force field $\propto r^{-5}$. These circular solutions were already known to Newton, see Corollary 1 to Proposition 7 in Book 1 of (Newton, 1687). Another force law considered by Newton is $F \propto r^{-3}$, which is the only power law for which the duality breaks down. As shown in Proposition 9 of (Newton, 1687, Book 1), this is the power law for which the solutions are logarithmic spirals.

Exercises

8.1 Use the fact that, in the hyperbolic case, the energy circle $\{v^2 = 2h\}$ intersects the velocity circle C orthogonally (see Lemma 8.6) to show that the subset $C \cap \{v^2 > 2h\}$ of C is characterised by the condition $1 + e\cos\theta > 0$. Since, by the proof of Corollary 3.3, this range of θ corresponds to the solution of the Kepler problem, we see again that this subset constitutes the hodograph.

8.2 A **k-dimensional sphere** in $\mathbb{R}^n$, $0 \le k \le n-1$, with centre $\mathbf{o}$ and radius r is a set of the form

$$\left\{\mathbf{x} = \mathbf{o} + \sum_{i=0}^{k} \lambda_i \mathbf{v}_i \in \mathbb{R}^n :\ \sum_{i=0}^{k} \lambda_i^2 = r^2\right\};$$

here $\mathbf{o}$ is an arbitrary point in $\mathbb{R}^n$, and $\mathbf{v}_0, \ldots, \mathbf{v}_k$ a $(k+1)$-tuple of orthonormal vectors. Show the following.

(a) For $k = n-1$ this definition amounts to the same as

$$\{\mathbf{x} \in \mathbb{R}^n :\ |\mathbf{x} - \mathbf{o}| = r\}.$$

(b) Every k-dimensional sphere is the intersection of an $(n-1)$-dimensional sphere (with the same centre $\mathbf{o}$) and an affine subspace of $\mathbb{R}^n$ containing $\mathbf{o}$.

(c) The intersection of two spheres of arbitrary dimension in $\mathbb{R}^n$ is a sphere, a point, or empty.

8.3 Let $\Psi\colon S^{n-1} \setminus \{N\} \to \mathbb{R}^{n-1}$ be the stereographic projection of the unit sphere $S^{n-1} \subset \mathbb{R}^n$ from the north pole $N := (0, \ldots, 0, 1) \in S^{n-1}$ to the equatorial plane $\{x_n = 0\} \equiv \mathbb{R}^{n-1}$. We wish to study the topological properties of the extended map $\Psi\colon S^{n-1} \to \mathbb{R}^{n-1} \cup \{\infty\}$ obtained by setting $\Psi(N) := \infty$.

We give $S^{n-1} \subset \mathbb{R}^n$ the **topology induced from** $\mathbb{R}^n$, i.e. the open sets in S^{n-1} are precisely the intersections $U \cap S^{n-1}$ with $U \subset \mathbb{R}^n$ open. The set $\mathbb{R}^{n-1} \cup \{\infty\}$ will be given the topology of the **one-point compactification**, i.e. the open subsets of $\mathbb{R}^{n-1} \cup \{\infty\}$ are said to be the sets of the following type:

(i) open subsets in $\mathbb{R}^{n-1}$;

(ii) sets of the form $(\mathbb{R}^{n-1} \setminus K) \cup \{\infty\}$, where $K \subset \mathbb{R}^{n-1}$ is a compact subset.

Show the following.

(a) This does indeed define a **topology** on $\mathbb{R}^{n-1} \cup \{\infty\}$, that is,

(1) $\emptyset$ and $\mathbb{R}^{n-1} \cup \{\infty\}$ are open;

(2) the intersection of finitely many open sets is open;

(3) the union of any number of open sets is open.

(b) The topological space $\mathbb{R}^{n-1} \cup \{\infty\}$ is **compact**, that is, given any covering of $\mathbb{R}^{n-1} \cup \{\infty\}$ by open sets, one can select finitely many of these sets that still cover the whole space.

(c) The map $\Psi\colon S^{n-1} \to \mathbb{R}^{n-1} \cup \{\infty\}$ is a **homeomorphism**, i.e. bijective and continuous in either direction, where **continuity** means that the preimage of any open set is open.

8.4 Let

$$\Psi_{\pm}\colon S^{n-1} \setminus \{(0, \ldots, 0, \pm 1)\} \longrightarrow \mathbb{R}^{n-1}$$

be the stereographic projection to the equatorial plane $\mathbb{R}^{n-1} \equiv \{x_n = 0\}$ from the north or south pole, respectively. Show that the composition of maps

$$\Psi_{-} \circ \Psi_{+}^{-1}\colon \mathbb{R}^{n-1} \setminus \{\mathbf{0}\} \longrightarrow \mathbb{R}^{n-1} \setminus \{\mathbf{0}\}$$

is given by $\mathbf{w} \mapsto \mathbf{w}/w^2$, i.e. it constitutes the inversion in the unit sphere $S^{n-2} \subset \mathbb{R}^{n-1}$.

Hint: This can be done by a computation, but it is much nicer to give a geometric proof. It suffices to consider the situation in the plane containing the poles $(0, \ldots, 0, \pm 1)$ and a point $\mathbf{x} \in S^{n-1}$, as well as the image points $\Psi_{\pm}(\mathbf{x})$. Then argue with similar triangles as in the proof of Lemma 8.44, for instance.

8.5 In this exercise we want to give a direct proof that stereographic projection is conformal. For simplicity, let us think of stereographic projection Ψ of $S^2 \subset \mathbb{R}^3$ from the north pole $N := (0, 0, 1)$ onto the equatorial plane $\mathbb{R}^2 \times \{0\}$. Prove conformality of Ψ at a given point $\mathbf{x} \in S^2 \setminus \{N\}$ by considering pairs of circles in S^2 that intersect at $\mathbf{x}$ and at N.

8.6 (a) Provide the details for the statement in the proof of Proposition 8.10 that by reversing the argument in the proof of Lemma 8.9 one can see that inversion in $\mathbb{S}_0$ sends spheres through $\mathbf{p}_0$ to planes not containing $\mathbf{p}_0$, and *vice versa*.

(b) Let now $\Phi\colon \mathbb{R}^3 \setminus \{\mathbf{0}\} \to \mathbb{R}^3 \setminus \{\mathbf{0}\}$ be the inversion in the unit sphere $S^2 \subset \mathbb{R}^3$ centred at the origin. Compute the Jacobian matrix $J_{\Phi}(\mathbf{x})$ of Φ and show that at every point $\mathbf{x} \in \mathbb{R}^3 \setminus \{\mathbf{0}\}$ it is proportional to an orthogonal matrix.

(c) Show that a C^1-map between open subsets of $\mathbb{R}^n$ is conformal if and only if its Jacobian matrix is proportional to an orthogonal matrix at every point in the domain of definition.

8.7 Show that the inverse map to the stereographic projection

$$\begin{array}{rccc} \Psi\colon & S^3 \setminus \{N\} & \longrightarrow & \mathbb{R}^3 \\ & (x_1, x_2, x_3, x_4) & \longmapsto & \dfrac{1}{1 - x_4}(x_1, x_2, x_3) \end{array}$$

onto the equatorial plane is given by

$$\Psi^{-1}(w_1, w_2, w_3) = \frac{1}{w^2 + 1}(2w_1, 2w_2, 2w_3, w^2 - 1),$$

where $w^2 = w_1^2 + w_2^2 + w_3^2$. This gives a direct proof that $\Psi\colon S^3 \setminus \{N\} \to \mathbb{R}^3$ is a diffeomorphism.

8.8 In this exercise we want to discuss geodesics on spheres. Given a submanifold $M \subset \mathbb{R}^n$, one can measure the length of a C^1-curve

$$\gamma\colon [a, b] \longrightarrow M \subset \mathbb{R}^n$$

in M simply as its length in $\mathbb{R}^n$ with respect to the euclidean norm:

$$\mathcal{L}(\gamma) := \int_a^b |\dot{\gamma}(t)| \, \mathrm{d}t.$$ [16]

Such a curve γ will be called a **geodesic** if it is parametrised proportionally to arc length, i.e. $|\dot{\gamma}|$ is a constant function, and if it is locally distance minimising. This means that for sufficiently small subintervals $[t_0, t_1] \subset [a, b]$, the curve $\gamma|_{[t_0,t_1]}$ is the shortest connection from $\gamma(t_0)$ to $\gamma(t_1)$.

(a) Now let M be the unit 2-sphere $S^2 \subset \mathbb{R}^3$ with the meridian

$$\{(x, y, z) \in S^2 \colon x \le 0, y = 0\}$$

removed. A point in M can be described in terms of its longitude φ with $-\pi < \varphi < \pi$ and its latitude $\pi/2 - \theta$ with $0 < \theta < \pi$. Draw a picture indicating these **spherical coordinates** (φ, θ) and show that the corresponding point on M has cartesian coordinates

$$(\sin\theta\cos\varphi, \sin\theta\sin\varphi, \cos\theta).$$

(b) Let $t \mapsto \gamma(t)$ be a C^1-curve on M, described in terms of spherical coordinates $(\theta(t), \varphi(t))$. Show that $|\dot{\gamma}(t)|^2 = \dot{\theta}^2 + \dot{\varphi}^2 \sin^2\theta$.

[16] If γ is an arbitrary C^1-curve, this integral measures the total distance travelled. If we want the integral to measure the geometric length of the trace $\gamma([a, b])$, the parametrisation has to be regular, i.e. $|\dot{\gamma}| > 0$, or at least the trace of the curve has to be traversed in a single direction only.

(c) We now want to show that arcs of great circles on S^2 are geodesics. Without loss of generality we consider an arc $\{\theta_0 < \theta < \theta_1, \varphi = 0\}$ on the prime meridian. Use (b) to show that the shortest curve joining the two end-points of this meridional arc is that very arc.

8.9 I claim that, alternatively, geodesics $\boldsymbol{\gamma}$ on a submanifold $M \subset \mathbb{R}^n$ can be characterised within the class of C^2-curves as those curves whose acceleration $\ddot{\boldsymbol{\gamma}}$ is always orthogonal to M, that is, $\ddot{\boldsymbol{\gamma}}(t)$ is orthogonal to the tangent space of M at the point $\mathbf{p} = \boldsymbol{\gamma}(t)$. This **tangent space** should be thought of as the vector space of velocity vectors at $\mathbf{p}$ of curves in M passing through $\mathbf{p}$, see Definition 8.21.[17]

Show that a curve satisfying this assumption on its acceleration is necessarily parametrised proportionally to arc length.

The aim of the remaining exercise is to prove one half of the claim above. Suppose we are given a *unit speed* C^2-curve

$$[a, b] \ni s \longmapsto \boldsymbol{\gamma}(s) \in M$$

with $\boldsymbol{\gamma}''$ not everywhere orthogonal to M. We want to show that $\boldsymbol{\gamma}$ is not distance minimising.

Without loss of generality we may assume that $\boldsymbol{\gamma}([a, b])$ is contained in the image $\mathbf{x}(U)$ of a smooth local parametrisation

$$\begin{array}{rccc} \mathbf{x}\colon & \mathbb{R}^m \supset U & \longrightarrow & M \subset \mathbb{R}^n \\ & (u^1, \dots, u^m) & \longmapsto & \mathbf{x}(u^1, \dots, u^m), \end{array}$$

where m is the dimension of the submanifold M. We set

$$\mathbf{x}_i := \frac{\partial \mathbf{x}}{\partial u^i}, \quad i = 1, \dots, m.$$

The curve $\boldsymbol{\gamma}$ can then be written as

$$\boldsymbol{\gamma}(s) = \mathbf{x}(\gamma_1(s), \dots, \gamma_m(s))$$

with C^2-functions $\gamma_i\colon\ [a, b] \to \mathbb{R}$, $i = 1, \dots, m$.

Let $\mathbf{v}(s)$ be the orthogonal projection of $\boldsymbol{\gamma}''(s)$ to the tangent space of M at the point $\boldsymbol{\gamma}(t)$. By assumption, $\mathbf{v}$ is not identically zero along $\boldsymbol{\gamma}$. Choose a smooth function $\lambda\colon\ [a, b] \to \mathbb{R}_0^+$ with $\lambda(a) = \lambda(b) = 0$ and $\lambda > 0$ on the open interval (a, b). Then write

$$\lambda(s)\mathbf{v}(s) = \sum_{i=1}^{m} v_i(s)\mathbf{x}_i(\gamma_1(s), \dots, \gamma_m(s))$$

[17] The restriction of the standard euclidean scalar product on $\mathbb{R}^n$ to the tangent spaces of M defines what is called a **Riemannian metric** on M, cf. Section 8.4.

with suitable functions v_i, and define

$$\alpha(\tau, s) := \mathbf{x}(\gamma_1(s) + \tau v_1(s), \ldots, \gamma_m(s) + \tau v_m(s))$$

for $s \in [a, b]$ and τ in a small interval around zero. Let $\mathcal{L}(\tau)$ be the length of the curve $s \mapsto \alpha(\tau, s)$ joining $\boldsymbol{\gamma}(a)$ and $\boldsymbol{\gamma}(b)$, i.e.

$$\mathcal{L}(\tau) = \int_a^b \left| \frac{\partial \alpha}{\partial s} \right| \mathrm{d}s.$$

(a) Show that

$$\frac{\mathrm{d}\mathcal{L}}{\mathrm{d}\tau} = \int_a^b \left\langle \frac{\partial^2 \alpha}{\partial s\, \partial \tau}, \frac{\partial \alpha}{\partial s} \right\rangle \cdot \left| \frac{\partial \alpha}{\partial s} \right|^{-1} \mathrm{d}s.$$

(b) Using the fact that $s \mapsto \alpha(0, s) = \boldsymbol{\gamma}(s)$ is a unit speed curve, and observing that

$$\left\langle \frac{\partial^2 \alpha}{\partial s\, \partial \tau}, \frac{\partial \alpha}{\partial s} \right\rangle = \frac{\mathrm{d}}{\mathrm{d}s} \left\langle \frac{\partial \alpha}{\partial \tau}, \frac{\partial \alpha}{\partial s} \right\rangle - \left\langle \frac{\partial \alpha}{\partial \tau}, \frac{\partial^2 \alpha}{\partial s^2} \right\rangle,$$

show that $\mathrm{d}\mathcal{L}/\mathrm{d}\tau\,(0) < 0$. This means that, by pushing $\boldsymbol{\gamma}$ in the direction of the tangential component of its acceleration, one can shorten the curve.

(c) Observe that this calculation also shows the following. If $\boldsymbol{\gamma}$ is a curve on M with acceleration everywhere orthogonal to M (and hence, without loss of generality, a unit speed curve), then $\mathrm{d}\mathcal{L}/\mathrm{d}\tau\,(0) = 0$ for any variation $\alpha(\tau, s)$ as above, where $\mathbf{v}$ is allowed to be any vector field along $\boldsymbol{\gamma}$ tangent to M.

8.10 Show that a unit speed C^2-curve $s \mapsto \boldsymbol{\gamma}(s)$ on the unit sphere $S^{n-1} \subset \mathbb{R}^n$ with $\boldsymbol{\gamma}''$ orthogonal to S^{n-1} lies on a great circle.

Hint: For $n = 3$ consider $\boldsymbol{\gamma} \times \boldsymbol{\gamma}'$. For $n > 3$ one may assume that $\boldsymbol{\gamma}(s_0), \boldsymbol{\gamma}'(s_0)$ are the unit vectors in the x_1- and x_2-direction, respectively. Consider the (x_1, x_2, x_i)-components of $\boldsymbol{\gamma}$ and $\boldsymbol{\gamma}'$ for $3 \leq i \leq n$, and then argue as for $n = 3$. Alternatively, from the identities $\langle \boldsymbol{\gamma}, \boldsymbol{\gamma} \rangle = 1$ and $\langle \boldsymbol{\gamma}', \boldsymbol{\gamma}' \rangle = 1$ you can show that $\boldsymbol{\gamma}'' = -\boldsymbol{\gamma}$; then use the Picard–Lindelöf theorem.

8.11 Draw a sketch of the great circles $\boldsymbol{\gamma}_\alpha$ from Proposition 8.17 on

$$S^2 \subset \mathbb{R}^3 \equiv \mathbb{R}^2 \times \{0\} \times \mathbb{R} \subset \mathbb{R}^4$$

and the corresponding velocity curves $\Psi \circ \boldsymbol{\gamma}_\alpha$ in

$$\mathbb{R}^2 \equiv \mathbb{R}^2 \times \{0\} \subset \mathbb{R}^3.$$

8.12 Let Δ be a **spherical triangle** on the unit sphere $S^2 \subset \mathbb{R}^3$, determined by three points and arcs of great circles between them. Write α, β, γ for the interior angles of Δ. Show by an elementary geometric argument that the area A of Δ satisfies $\alpha + \beta + \gamma = \pi + A$, see Figure 8.17.

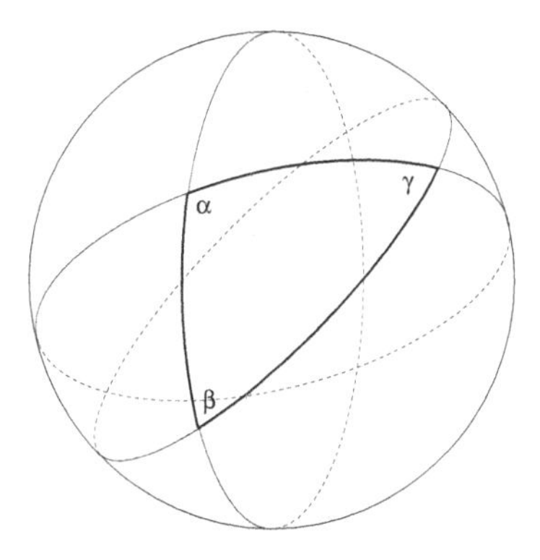

Figure 8.17 A spherical triangle.

8.13 (a) Show that for any point $(c_1, \ldots, c_n) \in \mathbb{H}$ the map

$$(x_1, \ldots, x_n) \longmapsto c_n(x_1, \ldots, x_n) + (c_1, \ldots, c_{n-1}, 0)$$

is an isometry of hyperbolic space $\mathbb{H}$.

(b) Conclude that the isometry group of $\mathbb{H}$, i.e. the set of all isometries of $\mathbb{H}$ with group multiplication given by the composition of maps, acts **transitively** on $\mathbb{H}$, which means that for any two points $\mathbf{x}, \mathbf{y} \in \mathbb{H}$ there is an isometry mapping $\mathbf{x}$ to $\mathbf{y}$.

(c) Now specialise to the case $n = 2$ and identify $\mathbb{H}$ with the upper half-plane $\{\operatorname{Im} \mathbf{z} > 0\}$ in $\mathbb{C}$. Show the following.

(1) For $a, b, c, d \in \mathbb{R}$ with $ad - bc = 1$, the fractional linear transformation

$$\mathbf{z} \longmapsto \frac{a\mathbf{z} + b}{c\mathbf{z} + d},$$

also known as a **Möbius transformation**, is an isometry of $\mathbb{H}$.

(2) The map

$$\mathbf{z} \longmapsto 1/\bar{\mathbf{z}}$$

is an isometry of $\mathbb{H}$.

(3) Any inversion in a circle of radius r about a point $\mathbf{p} \in \partial\mathbb{H}$, restricted to $\mathbb{H}$, can be written as a composition of maps as in (1) or (2).

(4) The maps as in (1) or (2) generate the full isometry group of $\mathbb{H}$.

Hint: Given an isometry of $\mathbb{H}$, show that by composing it with maps as in (1) or (2) one can obtain an isometry that fixes the points i and 2i $\in \mathbb{H}$, say, and preserves the orientation of $\mathbb{H}$. Explain why such an isometry must be the identity.

8.14 A k-dimensional submanifold of $\mathbb{H}$ will be called a **hyperbolic k-plane** if it contains the hyperbolic line through any two of its points. Show the following statements.

(a) The hyperbolic k-planes are the k-dimensional euclidean half-planes orthogonal to $\partial\mathbb{H}$ (we shall call these 'type 1') and the k-dimensional euclidean hemispheres orthogonal to $\mathbb{H}$ ('type 2').

(b) For any given three points in $\mathbb{H}$ there is a hyperbolic 2-plane containing the points.

(c) Any hyperbolic k-plane of type 2 can be mapped to a k-plane of type 1 by an isometry of $\mathbb{H}$.

8.15 Let ℓ be a hyperbolic line in the half-space model $\mathbb{H}$ of hyperbolic space. Construct a hyperbolic 2-plane that contains ℓ and the point $(0, \ldots, 0, 1)$. Observe that this hyperbolic 2-plane is mapped to a radial 2-disc in $\mathbb{D}$ under the isometry $\mathbb{H} \to \mathbb{D}$.

8.16 Let Δ by a **hyperbolic triangle** in the hyperbolic plane

$$\mathbb{H} = \{(x, y) \in \mathbb{R}^2 : \; y > 0\},$$

determined by three points and the geodesic arcs between them.

(a) Use the transformation formula for integrals to show that

$$A := \iint_\Delta \frac{\mathrm{d}x \, \mathrm{d}y}{y^2}$$

is invariant under isometries, i.e. for any hyperbolic isometry Φ we have

$$\iint_{\Phi(\Delta)} \frac{\mathrm{d}x \, \mathrm{d}y}{y^2} = \iint_\Delta \frac{\mathrm{d}x \, \mathrm{d}y}{y^2}.$$

We call A the **hyperbolic area** of Δ.

(b) By (a) we may assume without loss of generality that one of the sides of Δ is vertical. The other two sides of Δ can be parametrised in the form $\theta \mapsto (x_0 + a\cos\theta, a\sin\theta)$. By the theorem of Stokes we have

$$A = \int_{\partial\Delta} \frac{\mathrm{d}x}{y}.$$

Use this to show that $\alpha + \beta + \gamma = \pi - A$, where α, β, γ are the interior angles of Δ, cf. Exercise 8.12.[18]

8.17 Show that for regularised solutions of the Kepler problem with $h = 0$ and $c = 0$ the integral formula

$$u = \sqrt{\mu} \int_{t_0}^{t} \frac{d\tau}{r(\tau)}$$

holds for all $t \in \mathbb{R}$, even though the integrand is not defined for the collision time $t = t_0$.

Hint: Look at the solution $u \mapsto \mathbf{r}(u) = (-u^2/2, 0)$ in the proof of Proposition 8.34. Reparametrise this in terms of t and then compute the improper integral explicitly.

8.18 Verify the formulae on page 129.

8.19 Carry out the computation in the proof of Proposition 8.38.

8.20 Show that for $K > 0$ the map $\mathbf{w} \mapsto \sqrt{K}\,\mathbf{w}$ defines an isometry from the space form M_K (Definition 8.39) to the unit sphere S^3 with the Riemannian metric

$$\frac{4\langle\,.\,,\,.\,\rangle}{K(1+w^2)^2}.$$

Likewise, for $K < 0$ the map $\mathbf{w} \mapsto \sqrt{|K|}\,\mathbf{w}$ defines an isometry from M_K to the unit disc $\mathbb{D}^3$ with the metric

$$\frac{4\langle\,.\,,\,.\,\rangle}{|K|(1-w^2)^2}.$$

8.21 Here we give an alternative proof of Lemma 8.44. Let $a \subset \mathbb{R}^2$ be a line not passing through the origin O. Invert a in the circle C_0 of radius 1 about O to obtain a circle through O. Now use the theorem of Thales to argue that the polars of the points on a are precisely the lines passing through the pole of a.

8.22 The purpose of this exercise is to derive the duality statement of Proposition 8.50 more geometrically from the duality between poles and polars expressed in Lemma 8.44. Starting from a planar C^2-curve $\boldsymbol{\alpha}$, we constructed the polar reciprocal $\boldsymbol{\alpha}^*$ as the curve of poles of the tangents of $\boldsymbol{\alpha}$.

[18] The condition $\alpha + \beta + \gamma = \pi + KA$ for the area A of any geodesic triangle with interior angles α, β, γ, where $K \in \mathbb{R}$ is some constant, characterises a space of constant sectional curvature K.

Let a_0 be the tangent line to α at some point $B_0 \in \alpha$, and let $A_0 \in \alpha^*$ be its pole. Consider the secant of α^* through A_0 and a second point $A_1 \in \alpha^*$, the pole of a tangent line a_1 at $B_1 \in \alpha$, see Figure 8.18. By Lemma 8.44, the pole C of the line A_0A_1 lies on the polars of both A_0 and A_1, i.e. in the intersection $a_0 \cap a_1$.

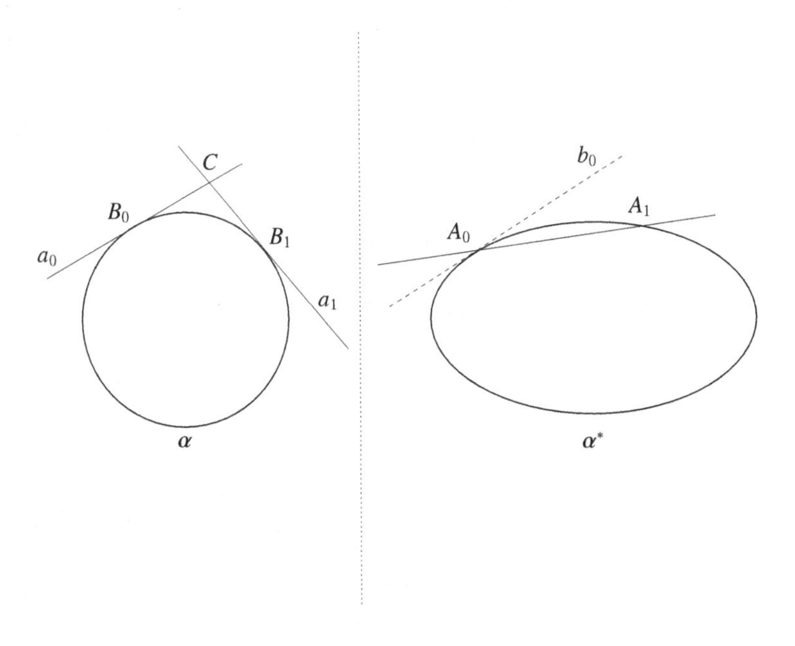

Figure 8.18 A synthetic proof of Proposition 8.50.

(a) Show that the condition that α have nowhere vanishing curvature guarantees that $a_0 \cap a_1$ is a single point C for A_1 near A_0, and that $C \to B_0$ for $A_1 \to A_0$.

(b) Use Lemma 8.44 to conclude that the secant A_0A_1 converges to the polar of B_0. So the tangent line b_0 of α^* at A_0 is defined, and its pole equals $B_0 \in \alpha$. This is again the statement of Proposition 8.50.

8.23 In the proof of Proposition 8.50 we saw that the angular momentum c^* of the polar reciprocal α^* of α vanishes precisely at the points corresponding to those points of α where the curvature κ vanishes. Here we wish to give a geometric explanation of this fact. Suppose, for instance, that $\alpha\colon (-\varepsilon, \varepsilon) \to \mathbb{R}^2$ has $\kappa > 0$ for $t < 0$ and $\kappa < 0$ for $t > 0$. This means that the tangents to α turn left for $t < 0$, and right for $t > 0$. Thus, every tangent direction near $\dot{\alpha}(0)$ is realised for a negative and for a positive value of t near 0. Deduce from the geometric definition of the polar reciprocal that $(\mathrm{d}\alpha^*/\mathrm{d}t)(0)$ points in the radial direction, which implies $c^*(0) = 0$.

8.24 Consider a family of lines b_t, $t \in I$, in $\mathbb{R}^2$, given by a parametric family of equations

$$a_0(t) + a_1(t)\xi + a_2(t)\eta = 0,$$

where $a_1^2(t) + a_2^2(t) \neq 0$ for all $t \in I$. An **envelope** of this set of lines is a C^1-curve $t \mapsto \boldsymbol{\beta}(t)$ such that b_t is the tangent line to $\boldsymbol{\beta}$ at $\boldsymbol{\beta}(t)$ for each $t \in I$. If we write $\boldsymbol{\beta}(t) = (\xi(t), \eta(t))$, this condition translates into

$$\left.\begin{array}{rcl} a_0(t) + a_1(t)\xi(t) + a_2(t)\eta(t) & = & 0 \\ a_1(t)\dot{\xi}(t) + a_2(t)\dot{\eta}(t) & = & 0. \end{array}\right\}$$

(a) Show that, provided the coefficient functions a_i are of class C^1, this system of equations is equivalent to

$$\left.\begin{array}{rcl} a_0 + a_1\xi + a_2\eta & = & 0 \\ \dot{a}_0 + \dot{a}_1\xi + \dot{a}_2\eta & = & 0. \end{array}\right\}$$

(b) Conclude that the envelope exists and is unique if $a_1\dot{a}_2 - a_2\dot{a}_1$ does not have any zeros.

(c) Let $t \mapsto \alpha(t) = (x(t), y(t))$, $t \in I$, be a planar C^1-curve with nowhere vanishing angular momentum. Show that the polar b_t of the point $\alpha(t)$ is the line

$$-1 + x(t)\xi + y(t)\eta = 0.$$

(d) Compute the envelope of this family of polars, and observe that it coincides with the polar reciprocal $\boldsymbol{\alpha}^*$ of $\boldsymbol{\alpha}$. This is a further manifestation of the duality expressed in Lemma 8.44.

(e) Show that a C^2-curve with nowhere vanishing curvature is the unique envelope of its family of tangent lines.

8.25 Let $I \ni t \mapsto \mathbf{r}(t) \in \mathbb{R}^2 \equiv \mathbb{C}$ be a regular C^2-curve with curvature function κ. Let $f\colon U \to \mathbb{C}$ be a holomorphic map whose derivative is nowhere zero, defined on some open neighbourhood U of $\mathbf{r}(I)$. Show that the curvature function $\tilde{\kappa}$ of $f \circ \mathbf{r}$ is given by

$$\tilde{\kappa} = \frac{1}{|f'|} \cdot \left(\kappa + \operatorname{Im}\left(\frac{f''}{f'} \cdot \frac{\dot{\mathbf{r}}}{|\dot{\mathbf{r}}|}\right)\right).$$

Here the derivatives of f are meant to be evaluated at the point $\mathbf{r}(t)$.

Hint: Think of the numerator of κ as the z-component of $\dot{\mathbf{r}} \times \ddot{\mathbf{r}}$ and observe, as in the proof of Proposition 8.60, that for

$$\mathbf{a}, \mathbf{b}, \mathbf{c} \in \mathbb{C} \equiv \mathbb{R}^2 \times \{0\} \subset \mathbb{R}^3$$

we have

$$\mathbf{ca} \times \mathbf{cb} = |\mathbf{c}|^2 \cdot \mathbf{a} \times \mathbf{b},$$

since multiplication by a non-zero complex number is an orientation-preserving conformal map.

8.26 In this exercise we give an analytic proof of Proposition 8.51. It suffices to consider a parametrised conic section of the form

$$\mathbf{r}(t) = \frac{c^2/\mu}{1 + e\cos f(t)} (\cos f(t), \sin f(t)) =: (x(t), y(t))$$

with

$$\dot{f} = \frac{c}{r^2} = \frac{\mu^2}{c^3} (1 + e\cos f)^2$$

as in the Kepler problem. Then the angular momentum is constant and equal to c. Show that

$$\dot{x} = -\frac{\mu}{c} \sin f, \quad \dot{y} = \frac{\mu}{c} (e + \cos f).$$

This implies that the polar reciprocal $\mathbf{r}^*$ of $\mathbf{r}$ is given by

$$\mathbf{r}^*(t) = \frac{\mu}{c^2} (e + \cos f(t), \sin f(t)),$$

which lies on a circle of radius μ/c^2 about the point $(e\mu/c^2, 0)$. Notice that the origin lies inside the circle for $e < 1$, on the circle for $e = 1$, and outside the circle for $e > 1$.

8.27 In this exercise we want to prove the theorem from (Hamilton, 1847) that "the Newtonian law may be *characterized* as being the *Law of the Circular Hodograph*." Let $t \mapsto \mathbf{r}(t) = (x(t), y(t))$ be a solution of the central force problem $\ddot{\mathbf{r}} = -f(\mathbf{r}) \cdot \mathbf{r}/r$. Compute the curvature of the velocity curve $t \mapsto (X(t), Y(t)) := (\dot{x}(t), \dot{y}(t))$ and observe that this curvature is constant if and only if $f \propto r^{-2}$.

8.28 This exercise shows how to give a geometric proof of the parabolic case in Proposition 8.51. Let $\mathcal{P} \subset \mathbb{R}^2$ be a parabola with directrix $\{x = -d\}$ and focus at the origin O, so that $\mathcal{P}$ is described by the equation $(x+d)^2 = x^2 + y^2$, or $y^2 = d^2 + 2dx$.

(a) Show that $\mathcal{P}$ has the following *reflection property*[19]: a light ray coming in from the right along a horizontal line $\{y = \text{const.}\}$ will be reflected towards the focus of the parabola. To this end, show that the angle between the tangent line a to $\mathcal{P}$ at a point $P = (x_0, y_0) \in \mathcal{P}$ and the horizontal line $\{y = y_0\}$ equals the angle between a and PO.

[19] Cf. Exercise 3.10.

(b) Show that the foot A' of the perpendicular from O to a moves along the straight line $\{x = -d/2\}$ as P moves along $\mathcal{P}$.

(c) Conclude that the polar reciprocal of $\mathcal{P}$ is a circle passing through O.

8.29 Give a geometric proof of the hyperbolic case in Proposition 8.51. Again it may help first to derive a reflection property of hyperbolas. (Consider light rays emanating from one focus. What is their relation to the other focus after reflection?)

8.30 Verify that the quotient topology as defined (in the case of $\mathbb{RP}^2$) in Section 8.6.5 satisfies the axioms of a topology (see Exercise 8.3).

8.31 Show that the projective line $\mathbb{RP}^1$ is homeomorphic to the circle S^1.

8.32 Let

$$b = \{(v_1, v_2) \in \mathbb{R}^2 : \; -b_0 + v_1 b_1 + v_2 b_2 = 0\}$$

be a line in $\mathbb{R}^2$, where we may assume that $b_1^2 + b_2^2 = 1$. Show the following.

(a) Provided $O \notin b$, the pole B of b is given by $B = (b_1, b_2)/b_0$, which under the inclusion $\mathbb{R}^2 \subset \mathbb{RP}^2$ is mapped to $[b_0, b_1, b_2]$.

(b) Under the inclusion $\mathbb{R}^2 \subset (\mathbb{RP}^2)^*$, the extension of b to a projective line in $(\mathbb{RP}^2)^*$ is given by

$$\{(u_0, u_1, u_2) \in \mathbb{R}^3 : \; u_0 b_0 + u_1 b_1 + u_2 b_2 = 0\}.$$

So $B \in \mathbb{RP}^2$ is simply the point defining the line $b \subset (\mathbb{RP}^2)^*$.

8.33 Show that the general solution of the Hooke problem (H) in Section 8.7 for $\lambda < 0$ is, up to inessential choices, as claimed in equation (H_-).

Hint: Consider the point on the solution curve where the distance from the origin is minimal.

8.34 Compute the energy h_H of the solutions of the Hooke problem (H) with vanishing angular momentum, see Remark 8.61. Show that a non-constant solution satisfies $h_\mathrm{H} > 0$ precisely if it passes through the origin. Compare the solutions of the Kepler problem obtained from the latter solutions of (H) with the regularised solutions discussed in Sections 8.3 and 8.5.

8.35 Deduce Kepler's first law in the case $c_\mathrm{K} \neq 0$ and $h_\mathrm{K} \geq 0$ from Proposition 8.60. Notice a subtlety in the case $h_\mathrm{K} > 0$, corresponding to $\lambda < 0$. In this case, there are solutions of (H) with $h_\mathrm{H} < 0$; these give rise to solutions of (K) with $\mu < 0$, i.e. a repellent force centre.

8.36 Let $s \mapsto \mathbf{w}(s) \in \mathbb{C} \setminus \{\mathbf{0}\}$ be a solution of the equation

$$\mathbf{w}'' = -\lambda |\mathbf{w}|^a \cdot \frac{\mathbf{w}}{|\mathbf{w}|},$$

where a is a real parameter. This describes the motion in a conservative force field whose potential, for $a \neq -1$, is given by

$$U(\mathbf{w}) = \frac{\lambda}{a+1}|\mathbf{w}|^{a+1}.$$

Hence, by Proposition 1.8, the total energy

$$h_{\mathbf{w}} = \frac{1}{2}|\mathbf{w}'|^2 + U(\mathbf{w})$$

is a constant of motion.

For $a \neq -3$, set $\alpha := (a+3)/2$, and define A by the equation

$$(a+3)(A+3) = 4. \tag{8.14}$$

Define $t = t(s)$, up to an additive constant, by

$$\frac{\mathrm{d}t}{\mathrm{d}s} = |\mathbf{w}(s)|^{2(\alpha-1)} = |\mathbf{w}(s)|^{a+1},$$

and set $\mathbf{z}(t) = \mathbf{w}^{\alpha}(s(t))$. From now on we exclude the trivial case $\alpha = 1$, corresponding to $(a, A) = (-1, -1)$.

(a) Prove, generalising Proposition 8.60, that $\mathbf{z}$ is a solution of the central force problem

$$\ddot{\mathbf{z}} = -2\alpha(\alpha - 1)h_{\mathbf{w}}|\mathbf{z}|^{A} \cdot \frac{\mathbf{z}}{|\mathbf{z}|}.$$

Observe that the integer solutions of (8.14), excluding $(-1, -1)$, are given by

$$(a, A) \in \{(1, -2), (-2, 1), (-4, -7), (-7, -4), (-5, -5)\}.$$

(b) Show that the zero energy solutions for the force field $F \propto r^{-5}$ are circles passing through the origin.

8.37 The purpose of this exercise is to give a more direct derivation of Newton's result that the force required to keep a body on a circle through the force centre is proportional to r^{-5}.

Let $t \mapsto \mathbf{r}(t) \in \mathbb{R}^2 \setminus \{\mathbf{0}\} \subset \mathbb{R}^3$ be a solution to a given central force problem $\ddot{\mathbf{r}} = \mathbf{F}(\mathbf{r})$ (for a body of unit mass) with (constant) angular momentum $\mathbf{c} = (0, 0, c)$ with $c > 0$. (In particular, the curve $\mathbf{r}$ must then be regular.) Write $\gamma(t) \in [0, \pi/2)$ for the acute angle between $\mathbf{r}(t)$ and the normal to the curve. Let κ be the curvature function of the planar curve $\mathbf{r}$.

(a) We shall assume throughout that the force field is attracting, i.e. the force $\mathbf{F}$ is directed from $\mathbf{r}$ towards $\mathbf{0}$. Show that, under the assumption on $\mathbf{c}$, the angle from $\mathbf{v}$ to $\mathbf{F}$ is $\pi/2 \pm \gamma \in (0, \pi)$. Deduce from the definition of κ in Lemma 8.47 that $\kappa > 0$.

(b) Show that $F = \kappa v^2/\cos\gamma$ and $c = rv\cos\gamma$, where $v = |\dot{\mathbf{r}}|$. It follows that

$$F = c^2 \cdot \frac{\kappa}{r^2\cos^3\gamma},$$

which is essentially Proposition 7 in Book 1 of (Newton, 1687) translated into modern language.

(c) Now suppose $\mathbf{r}$ describes a circle of radius R through the origin. Draw a picture of this situation and show that $\cos\gamma = r/2R$. Deduce with $\kappa = 1/R$ that $F = 8R^2c^2r^{-5}$.

(d) Show that the energy $h = v^2/2 - \lambda/4r^4$ of the solutions in (c), where $\lambda = 8R^2c^2$, equals zero, cf. Exercise 8.36 (b).

8.38 In this exercise we show how one can base a more geometric proof of the duality in Exercise 8.36 on part (b) of the preceding exercise and the transformation formula for the curvature in Exercise 8.25. For simplicity, we deal only with the case $(a, A, \alpha) = (1, -2, 2)$ of Proposition 8.60.

Thus, let $\mathbf{w}$ be a solution of (H) with non-vanishing angular momentum, and let $f\colon \mathbb{C} \to \mathbb{C}$ be the map $\mathbf{w} \mapsto \mathbf{w}^2 =: \mathbf{z}$. Use the notation $\kappa, F, c, \gamma, r = |\mathbf{w}|$ from Exercise 8.37 for the quantities associated with $\mathbf{w}$, and $\tilde{\kappa}, \tilde{F}, \tilde{c}, \tilde{\gamma}, \tilde{r} = |\mathbf{z}|$ for those associated with $\mathbf{z}$. We regard F as negative in the case $\lambda < 0$, i.e. a repelling force; this corresponds to negative values of κ.

(a) Observe that $\tilde{\gamma} = \gamma$, since the ray from $\mathbf{0}$ to $\mathbf{w}$ is sent by the conformal map f to the ray from $\mathbf{0}$ to $\mathbf{z}$.

(b) Using Exercise 8.25, show that

$$\tilde{\kappa} = \frac{\kappa}{2r} + \frac{\cos\gamma}{2r^2}.$$

(c) Show with Exercise 8.37 that the energy $h = h_H$ of $\mathbf{w}$ can be expressed as

$$h = \frac{c^2}{\cos^3\gamma} \cdot \left(\frac{\kappa}{2r} + \frac{\cos\gamma}{2r^2}\right).$$

(d) Conclude that

$$\tilde{F} = h\frac{\tilde{c}^2}{c^2} \cdot \frac{1}{\tilde{r}^2}.$$

Notice that the dynamics is now hidden in c and $\tilde{c}$; with the parametrisations as in Proposition 8.60 we have $\tilde{c} = 2c$, and we recover the result $\mu = 4h$ from that proposition.

9

Hamiltonian mechanics

> Et comme dans les mathématiques, quand il n'y a point de *maximum* ni de *minimum*, rien enfin de distingué, tout se fait également; ou quand cela ne se peut, il ne se fait rien du tout; on peut dire de même en matière de parfaite sagesse, qui n'est pas moins réglée que les mathématiques, que s'il n'y avait pas le meilleur *(optimum)* parmi tous les mondes possibles, Dieu n'en aurait produit aucun.
>
> Gottfried Wilhelm Leibniz, *Essais de Théodicée*

Recall the equation (PCR3B) from Section 7.3 describing the planar circular restricted three-body problem, written in cartesian coordinates x, y (see page 87):

$$\left.\begin{aligned} \ddot{x} - 2\dot{y} &= \frac{\partial \Phi}{\partial x} \\ \ddot{y} + 2\dot{x} &= \frac{\partial \Phi}{\partial y}. \end{aligned}\right\}$$

This can be rewritten as

$$\left.\begin{aligned} \frac{\mathrm{d}}{\mathrm{d}t}(\dot{x} - y) &= \dot{y} + \frac{\partial \Phi}{\partial x} \\ \frac{\mathrm{d}}{\mathrm{d}t}(\dot{y} + x) &= -\dot{x} + \frac{\partial \Phi}{\partial y}. \end{aligned}\right\}$$

Set $q_1 = x$, $q_2 = y$, $p_1 = \dot{x} - y$, $p_2 = \dot{y} + x$. Then

$$\left.\begin{aligned} \dot{q}_1 &= p_1 + q_2 \\ \dot{q}_2 &= p_2 - q_1, \end{aligned}\right\} \tag{9.1}$$

and the system above becomes

$$\left.\begin{aligned} \dot{p}_1 &= p_2 - q_1 + \frac{\partial \Phi}{\partial q_1} \\ \dot{p}_2 &= -p_1 - q_2 + \frac{\partial \Phi}{\partial q_2}. \end{aligned}\right\} \tag{9.2}$$

Set

$$H(q_1, q_2, p_1, p_2) := \frac{1}{2}(p_1 + q_2)^2 + \frac{1}{2}(p_2 - q_1)^2 - \Phi(q_1, q_2). \tag{9.3}$$

Then the system of equations (9.1) and (9.2) can be written as follows:

$$\left.\begin{aligned} \dot{q}_1 &= \frac{\partial H}{\partial p_1} \quad , \quad \dot{q}_2 = \frac{\partial H}{\partial p_2} \\ \dot{p}_1 &= -\frac{\partial H}{\partial q_1} \quad , \quad \dot{p}_2 = -\frac{\partial H}{\partial q_2}. \end{aligned}\right\}$$

Equations of this type are called a **canonical system** or the **Hamilton equations** for the **Hamiltonian function** $H = H(\mathbf{q}, \mathbf{p})$.

This is the form in which one will typically find (PCR3B) described in the contemporary literature. The advantage is that we are now dealing with a particularly simple system of first-order equations.

Remark 9.1 When q_1, q_2, p_1, p_2 in H are substituted by their expressions in terms of $x, y, \dot{x}, \dot{y}$, the resulting function of t is $-\mathcal{J}/2$, where $\mathcal{J}$ is the Jacobi integral from Definition 7.3.

Section 9.1 contains a brief introduction to variational principles. In particular, I show that the Newtonian equation of motion can be derived as the Euler–Lagrange equation for the Lagrange functional (Hamilton's principle) or as the condition for a trajectory to be a critical point of the action functional (Maupertuis's principle). The so-called Jacobi metric resulting from the latter principle will then be used to give a beautiful proof of the duality of force laws encountered earlier in Section 8.7.

In Section 9.2 we see how a process known as Legendre transformation allows us to transform the Euler–Lagrange equation, which is of order two, into a first-order Hamiltonian system.

In Section 9.3 we study transformations that preserve the form of the Hamilton equations. These are known as 'canonical transformations' or, in modern parlance, 'symplectomorphisms'. This section requires knowledge of differential forms, and is not necessary for what follows. However, the elegant proof of energy conservation (Proposition 9.24) may serve to illustrate the power of this formalism. One of the arts in the theory of Hamiltonian systems is to find a canonical transformation of a given system into one that can be solved more or less explicitly. Such transformations are studied at length in (Meyer *et al.*, 2009).

Finally, in Section 9.4, we investigate the stability of the libration points in the planar circular restricted three-body problem.

9.1 Variational principles

In the famous opening sentence of (Siegel and Moser, 1971), the authors make the following remark: “Ours, according to Leibniz, is the best of all possible worlds, and the laws of nature can therefore be described in terms of extremal principles.” Whereas the premise from (Leibniz, 1710) has deservedly been lampooned by Voltaire (1759), the conclusion is amply confirmed by the plethora of geometric and physical problems that permit a variational formulation.

The basic general question in the calculus of variations asks us to minimise an expression of the form

$$\mathcal{S}(\alpha) = \int_a^b L(t, \alpha(t), \dot{\alpha}(t))\, \mathrm{d}t, \tag{9.4}$$

where the **Lagrangian function** L is a given C^2-function

$$\begin{array}{ccc} [a,b] \times \Omega \times \mathbb{R}^n & \longrightarrow & \mathbb{R} \\ (t, \mathbf{q}, \mathbf{v}) & \longmapsto & L(t, \mathbf{q}, \mathbf{v}), \end{array}$$

Ω is an open subset of $\mathbb{R}^n$, and α is sought among all C^2-functions $[a,b] \to \Omega$ with given values at a and b. For instance, finding the shortest curve (of class C^2) among all graphs (over x) connecting the points (a, q_a) and (b, q_b) in the xy-plane, i.e. a curve of the form $t \mapsto (t, \alpha(t))$, amounts to minimising the length functional

$$\mathcal{S}(\alpha) = \int_a^b \sqrt{1 + \dot{\alpha}^2}\, \mathrm{d}t,$$

given by integrating the speed along the curve, over all C^2-maps $\alpha\colon\ [a,b] \to \mathbb{R}$ with $\alpha(a) = q_a$ and $\alpha(b) = q_b$.

Here is the main theorem of this section.

Theorem 9.2 *If α minimises the functional $\mathcal{S}$ from* (9.4) *among all functions $\beta \in C^2([a,b], \Omega)$ with given $\beta(a) = \mathbf{q}_a$, $\beta(b) = \mathbf{q}_b$, then α satisfies the* **Euler–Lagrange equation**

$$\frac{\mathrm{d}}{\mathrm{d}t}\left(\frac{\partial L}{\partial \mathbf{v}}(t, \alpha, \dot{\alpha})\right) - \frac{\partial L}{\partial \mathbf{q}}(t, \alpha, \dot{\alpha}) = \mathbf{0} \ \textit{for all}\ \ t \in [a,b], \tag{EL}$$

where α and $\dot{\alpha}$ are meant to be read as functions of t.

Remark 9.3 What the proof below will show is that α satisfies the Euler–Lagrange equation if and only if it is a *critical* point of $\mathcal{S}$ with regard to variations of α that keep the end-points fixed.

Example 9.4 For the length functional described above, the Euler–Lagrange equation reads

$$\frac{\mathrm{d}}{\mathrm{d}t}\frac{\dot{\alpha}}{\sqrt{1+\dot{\alpha}^2}} = 0,$$

which is easily seen to be equivalent to $\ddot{\alpha} = 0$. This means that the only candidate for a minimal solution is, as expected, the straight line.

Example 9.5 Consider the motion of a particle of mass m in a force field $\mathbf{F}\colon \Omega \to \mathbb{R}^n$ given by a potential $V\colon \Omega \to \mathbb{R}$, that is, $\mathbf{F} = -\operatorname{grad} V$. Then the Euler–Lagrange equation for the Lagrangian function

$$L(\mathbf{q},\mathbf{v}) = \frac{1}{2}m|\mathbf{v}|^2 - V(\mathbf{q})$$

given by the *difference* of kinetic and potential energy is the Newtonian equation $m\ddot{\alpha} = -\operatorname{grad} V(\alpha)$. Indeed, we have

$$\frac{\partial L}{\partial \mathbf{v}} = m\mathbf{v},$$

and so the Euler–Lagrange equation becomes

$$\mathbf{0} = \frac{\mathrm{d}}{\mathrm{d}t}(m\dot{\alpha}) + \frac{\partial V}{\partial \mathbf{q}}(\alpha) = m\ddot{\alpha} + \operatorname{grad} V(\alpha).$$

The fact that motions of this mechanical system are precisely the critical points of the functional $\int L\,\mathrm{d}t$ is known as **Hamilton's principle**.

The key to proving Theorem 9.2 is the following **fundamental lemma of the calculus of variations.**

Lemma 9.6 *Let $f\colon [a,b] \to \mathbb{R}$ be a continuous function. If*

$$\int_a^b f(t)g(t)\,\mathrm{d}t = 0$$

for every $g \in C^2([a,b])$ with $g(a) = g(b) = 0$, then f is identically equal to zero.

Proof We prove the contrapositive statement. If the continuous function f is not identically equal to zero, there is a $t_0 \in (a,b)$ where $f(t_0) \neq 0$. Without loss of generality we may assume that $f(t_0)$ is positive. The continuity of f implies that there is a $\delta > 0$ such that $[t_0 - \delta, t_0 + \delta] \subset (a, b)$ and $f(t) > 0$ for all t in that subinterval.

There is a 'bump' function $g \in C^\infty([a, b]) \subset C^2([a, b])$ with

$$g(t) \begin{cases} = 0 & \text{for } t \in [a, t_0 - \delta] \text{ or } t \in [t_0 + \delta, b], \\ > 0 & \text{for } t \in (t_0 - \delta, t_0 + \delta). \end{cases}$$

Indeed, the function $\psi\colon \mathbb{R} \to \mathbb{R}$ defined by

$$\psi(t) := \begin{cases} \exp(-1/t) & \text{for } t > 0, \\ 0 & \text{for } t \leq 0, \end{cases}$$

is smooth, and we can set $g(t) = \psi(t - (t_0 - \delta)) \cdot \psi((t_0 + \delta) - t)$.

Then

$$\int_a^b f(t)g(t)\,\mathrm{d}t = \int_{t_0-\delta}^{t_0+\delta} f(t)g(t)\,\mathrm{d}t > 0. \qquad \square$$

Proof of Theorem 9.2 Let $\boldsymbol{\alpha}$ be a function with $\mathcal{S}(\boldsymbol{\alpha}) \leq \mathcal{S}(\boldsymbol{\beta})$ for all $\boldsymbol{\beta}$ within the class of allowable functions. Let $\mathbf{g} \in C^2([a, b], \mathbb{R}^n)$ be a function with $\mathbf{g}(a) = \mathbf{g}(b) = \mathbf{0}$. For τ in a sufficiently small interval around 0, we have $\boldsymbol{\alpha} + \tau\mathbf{g} \in C^2([a, b], \Omega)$, so we can define a real-valued function σ on that interval by $\sigma(\tau) := \mathcal{S}(\boldsymbol{\alpha} + \tau\mathbf{g})$. Then σ is differentiable and $\sigma(0) \leq \sigma(\tau)$ for all τ in the relevant interval around 0, which implies $\sigma'(0) = 0$. On the other hand,

$$\begin{aligned} \sigma'(\tau) &= \int_a^b \frac{\partial}{\partial \tau}(L(t, \boldsymbol{\alpha} + \tau\mathbf{g}, \dot{\boldsymbol{\alpha}} + \tau\dot{\mathbf{g}}))\,\mathrm{d}t \\ &= \int_a^b \left(\frac{\partial L}{\partial \mathbf{q}}\mathbf{g} + \frac{\partial L}{\partial \mathbf{v}}\dot{\mathbf{g}} \right) \mathrm{d}t, \end{aligned}$$

where the products in the last integrand are meant to be read as standard scalar products of vector-valued functions.

Through integration by parts we find

$$\int_a^b \frac{\partial L}{\partial \mathbf{v}}\dot{\mathbf{g}}\,\mathrm{d}t = \frac{\partial L}{\partial \mathbf{v}}\mathbf{g}\Big|_{t=a}^{t=b} - \int_a^b \frac{\mathrm{d}}{\mathrm{d}t}\left(\frac{\partial L}{\partial \mathbf{v}}\right)\mathbf{g}\,\mathrm{d}t = -\int_a^b \frac{\mathrm{d}}{\mathrm{d}t}\left(\frac{\partial L}{\partial \mathbf{v}}\right)\mathbf{g}\,\mathrm{d}t,$$

since $\mathbf{g}(a) = \mathbf{g}(b) = 0$. It follows that

$$0 = \sigma'(0) = \int_a^b \left[\frac{\partial L}{\partial \mathbf{q}}(t, \boldsymbol{\alpha}, \dot{\boldsymbol{\alpha}}) - \frac{\mathrm{d}}{\mathrm{d}t}\left(\frac{\partial L}{\partial \mathbf{v}}(t, \boldsymbol{\alpha}, \dot{\boldsymbol{\alpha}})\right) \right] \mathbf{g}\,\mathrm{d}t.$$

By choosing a vector-valued function $\mathbf{g}$ with only one component non-zero, we can apply the preceding lemma component-wise to deduce that the expression in brackets is identically zero. $\square$

Next we discuss geodesics in this variational context, cf. Section 8.4. Assume we are given a Riemannian metric on the open subset $\Omega \subset \mathbb{R}^n$, i.e. an inner product $\langle \, . \, , . \, \rangle_{\mathbf{q}}$ on the tangent space $T_{\mathbf{q}}\Omega \equiv \mathbb{R}^n$ (cf. Definition 8.21) varying smoothly with the point $\mathbf{q} \in \Omega$. This means that, if we express this inner product with respect to the standard basis of $\mathbb{R}^n$ as a positive-definite symmetric matrix, then the matrix entries should be smooth functions of $\mathbf{q}$.

For a C^1-curve $\gamma\colon [a, b] \to \Omega$ we then have the length functional

$$\mathcal{L}(\gamma) = \int_a^b |\dot{\gamma}(t)| \, \mathrm{d}t,$$

where $|\dot{\gamma}(t)|$ denotes the norm of the velocity vector $\dot{\gamma}(t) \in T_{\gamma(t)}\Omega$ with respect to the inner product at the point $\gamma(t)$.

Exercise 8.9 suggests the following definition.

Definition 9.7 A **geodesic** in Ω (with respect to the given Riemannian metric) is a critical point γ for the length functional $\mathcal{L}$ under variations with fixed endpoints (with γ and the variations of class C^2, and γ parametrised proportionally to arc length).

What Exercise 8.9 has shown is that on a submanifold $M \subset \mathbb{R}^k$ – with Riemannian metric given by restricting the euclidean inner product on $\mathbb{R}^k$ to the tangent spaces of M – the geodesics in the sense defined here (in a local chart of M) are precisely the curves with acceleration orthogonal to M.

Example 9.8 Given two non-antipodal points on $S^2 \subset \mathbb{R}^3$, there are two arcs of great circles joining them. Both are geodesics – also in the sense of being locally length minimising – but only the shorter of the two is a minimum of the length functional.

We now want to show from Definition 9.7 that the geodesics in the hyperbolic plane $\mathbb{H}$ are as described in Proposition 8.29. Rather than working directly from the Euler–Lagrange equation (EL), we derive a variant of this equation for **autonomous** Lagrangians, i.e. Lagrangian functions $L = L(\mathbf{q}, \mathbf{v})$ that do not depend explicitly on t.

Lemma 9.9 *If $L = L(\mathbf{q}, \mathbf{v})$ is autonomous, then every solution of* (EL) *satisfies the equation*

$$\dot{\alpha} \frac{\partial L}{\partial \mathbf{v}}(\alpha, \dot{\alpha}) - L(\alpha, \dot{\alpha}) = \text{const.} \tag{EL$'$}$$

Conversely, in dimension $n = 1$ any solution of (EL$'$) *with $\dot{\alpha} \neq 0$ is a solution of* (EL).

Proof We compute

$$\frac{\mathrm{d}}{\mathrm{d}t}\left(\dot{\alpha}\frac{\partial L}{\partial \mathbf{v}}(\alpha,\dot{\alpha}) - L(\alpha,\dot{\alpha})\right) = \ddot{\alpha}\frac{\partial L}{\partial \mathbf{v}} + \dot{\alpha}\frac{\mathrm{d}}{\mathrm{d}t}\left(\frac{\partial L}{\partial \mathbf{v}}(\alpha,\dot{\alpha})\right) - \frac{\partial L}{\partial \mathbf{q}}\dot{\alpha} - \frac{\partial L}{\partial \mathbf{v}}\ddot{\alpha}$$
$$= \dot{\alpha}\left[\frac{\mathrm{d}}{\mathrm{d}t}\left(\frac{\partial L}{\partial \mathbf{v}}(\alpha,\dot{\alpha})\right) - \frac{\partial L}{\partial \mathbf{q}}(\alpha,\dot{\alpha})\right].$$

From this identity the lemma follows. □

For instance, in the case of Example 9.5, the constant of motion provided by the left-hand side of (EL′) is simply the total energy of the particle (Exercise 9.1).

Example 9.10 We want to determine the geodesics in $\mathbb{H}$ that can be written as graphs over the x-axis, i.e. geodesics that can be parametrised as $t \mapsto (t, \alpha(t))$ over some interval $[a, b]$. For these, the length functional from Definition 8.18 becomes

$$\mathcal{L}(\alpha) = \int_a^b \frac{\sqrt{1 + \dot{\alpha}^2(t)}}{\alpha(t)}\,\mathrm{d}t.$$

In order to apply the preceding lemma, we also assume $\dot{\alpha} \neq 0$. Then a necessary and sufficient condition for $t \mapsto (t, \alpha(t))$ to be a geodesic (when reparametrised proportionally to arc length) is that

$$\dot{\alpha}\frac{\dot{\alpha}}{\alpha\sqrt{1+\dot{\alpha}^2}} - \frac{\sqrt{1+\dot{\alpha}^2}}{\alpha} = \frac{-1}{\alpha\sqrt{1+\dot{\alpha}^2}}$$

be constant. This condition characterises semicircles orthogonal to $\partial\mathbb{H}$ (Exercise 9.2).

Historically one of the most important incarnations of the variational principle is the least action principle of Maupertuis. As in Example 9.5 we study the motion $\boldsymbol{\alpha}$ of a particle of mass m in a conservative force field on $\Omega \subset \mathbb{R}^n$ given by a potential $V\colon \Omega \to \mathbb{R}$. We only consider motions $\boldsymbol{\alpha}$ for a fixed value h of the energy

$$\frac{1}{2}m|\dot{\boldsymbol{\alpha}}|^2 + V(\boldsymbol{\alpha}).$$

This means that the speed $v = |\dot{\boldsymbol{\alpha}}|$ becomes a function of the position:

$$v = \sqrt{\frac{2(h - V(\mathbf{q}))}{m}}. \tag{9.5}$$

Motion is possible only in **Hill's region**

$$\{\mathbf{q} \in \Omega\colon\ h - V(\mathbf{q}) \geq 0\},$$

cf. Definition 7.7. For curves α in that region, we can define the **action functional**

$$\mathcal{A}(\alpha) = \int_\alpha mv\,\mathrm{d}s,$$

that is, we integrate v given by (9.5) along the curve with respect to the arc length element $\mathrm{d}s$ for the euclidean metric. The value of this integral does not depend on the choice of (regular) parametrisation $t \mapsto \alpha(t)$. The action can be computed with the arc length element

$$\mathrm{d}s = |\dot{\alpha}(t)|\,\mathrm{d}t$$

and v given by (9.5) for any such parametrisation, even if $v(\alpha(t)) \neq |\dot{\alpha}(t)|$. Thus, for the computation of the critical points of the action functional $\mathcal{A}$, the choice of parametrisation is irrelevant.

Theorem 9.11 (Maupertuis's principle) *Among C^2-curves in Hill's region joining two given points, critical points for the action functional $\mathcal{A}$, when parametrised such that the speed conforms to equation* (9.5)*, are precisely the trajectories of the Newtonian equation*

$$m\ddot{\mathbf{q}} = -\operatorname{grad} V(\mathbf{q}). \qquad (\mathrm{N_c})$$

of energy h.

The expression

$$mv\,\mathrm{d}s = \sqrt{2m(h - V(\mathbf{q}))}\,\mathrm{d}s$$

can be interpreted as the arc length element for the rescaled metric

$$2m(h - V(\mathbf{q}))\,\langle\,.\,,\,.\,\rangle$$

on Hill's region, called the **Jacobi metric**. So the following is evident.

Corollary 9.12 *Trajectories α of the Newtonian equation* ($\mathrm{N_c}$) *of energy h, when parametrised such that*

$$2m(h - V(\alpha))\,|\dot{\alpha}|^2 = \text{const.},$$

are precisely the geodesics for the Jacobi metric. □

Proof of Theorem 9.11 Consider regular C^2-curves joining two given points in Hill's region. To start with, we take an arbitrary parametrisation of these curves on the interval $[a, b]$. The action functional can then be written as

$$\mathcal{A}(\alpha) = \int_a^b \sqrt{2m(h - V(\alpha))}\,|\dot{\alpha}|\,\mathrm{d}t.$$

So we are in the setting of Theorem 9.2 with

$$L(\mathbf{q}, \mathbf{v}) := \sqrt{2m(h - V(\mathbf{q}))}\,|\mathbf{v}|.$$

The Euler–Lagrange equation becomes

$$\frac{\mathrm{d}}{\mathrm{d}t}\left(\sqrt{2m(h - V(\boldsymbol{\alpha}))}\,\frac{\dot{\boldsymbol{\alpha}}}{|\dot{\boldsymbol{\alpha}}|}\right) + \sqrt{\frac{m}{2(h - V(\boldsymbol{\alpha}))}}\,\frac{\partial V}{\partial \mathbf{q}}(\boldsymbol{\alpha})\,|\dot{\boldsymbol{\alpha}}| = \mathbf{0}.$$

Since the action functional is independent of the parametrisation of $\boldsymbol{\alpha}$, this Euler–Lagrange equation will hold for a critical point $\boldsymbol{\alpha}$ no matter which parametrisation we choose, so let it be the one satisfying (9.5), that is,

$$|\dot{\boldsymbol{\alpha}}| = \sqrt{\frac{2(h - V(\boldsymbol{\alpha}))}{m}},$$

which makes $\boldsymbol{\alpha}$ of constant energy h. Then the Euler–Lagrange equation becomes

$$m\ddot{\boldsymbol{\alpha}} + \frac{\partial V}{\partial \mathbf{q}}(\boldsymbol{\alpha}) = \mathbf{0},$$

which is the Newtonian equation ($\mathrm{N_c}$). □

Corollary 9.12 allows us to give a quick and conceptually much nicer proof of Proposition 8.60 and its generalisations.

Theorem 9.13 *Let $\mathbf{w} \mapsto \mathbf{z}$ be a biholomorphic mapping between certain open subsets of $\mathbb{C}$. Under this mapping, the trajectories of energy h for the potential*

$$U(\mathbf{w}) = \left|\frac{\mathrm{d}\mathbf{z}}{\mathrm{d}\mathbf{w}}\right|^2,$$

that is, solutions of

$$\mathbf{w}'' = -\frac{\partial U}{\partial \mathbf{w}}(\mathbf{w})$$

with $|\mathbf{w}|^2/2 + U(\mathbf{w}) = h$ (notice that $h > 0$), transform into trajectories of energy $-1/h$ for the potential

$$V(\mathbf{z}) = -\left|\frac{\mathrm{d}\mathbf{w}}{\mathrm{d}\mathbf{z}}\right|^2,$$

that is, solutions of

$$\ddot{\mathbf{z}} = -\frac{\partial V}{\partial \mathbf{z}}(\mathbf{z})$$

with $|\mathbf{z}|^2/2 + V(\mathbf{z}) = -1/h$.

Proof This follows from Corollary 9.12 and the simple calculation

$$(h-U)\,|\mathrm{d}\mathbf{w}|^2 = \left(h\left|\frac{\mathrm{d}\mathbf{w}}{\mathrm{d}\mathbf{z}}\right|^2 - 1\right)|\mathrm{d}\mathbf{z}|^2 = h\left(-\frac{1}{h} - V\right)|\mathrm{d}\mathbf{z}|^2. \qquad \square$$

Remark 9.14 If you are a little worried about the 'Leibnizian' notation in that proof (and I think you should be), here is how to translate it into that of Chapter 8. Write the biholomorphism $\mathbf{w} \mapsto \mathbf{z}$ as φ. The length element $|\mathrm{d}\mathbf{z}|$ measures the euclidean length of a tangent vector $\mathbf{v} \in T_{\mathbf{z}}\mathbb{C}$, that is, $|\mathrm{d}\mathbf{z}|^2(\mathbf{v}) = \langle \mathbf{v}, \mathbf{v}\rangle$. By polarisation, see (8.7), the length element determines the metric.

As is habitual in Riemannian geometry, let us write g for a given Riemannian metric on the domain of the variable $\mathbf{z}$, that is, the inner product of $\mathbf{v}, \mathbf{v}' \in T_{\mathbf{z}}\mathbb{C}$ is written as $g_{\mathbf{z}}(\mathbf{v}, \mathbf{v}')$.

The transformation φ allows us to pull back this metric to a metric $\varphi^* g$ on the domain of the variable $\mathbf{w}$ by setting

$$(\varphi^* g)_{\mathbf{w}}(\mathbf{u}, \mathbf{u}') := g_{\varphi(\mathbf{w})}(J_{\varphi,\mathbf{w}}(\mathbf{u}), J_{\varphi,\mathbf{w}}(\mathbf{u}')) \quad \text{for } \mathbf{u}, \mathbf{u}' \in T_{\mathbf{w}}\mathbb{C}.$$

In other words, $\varphi^* g$ is the metric for which the map φ becomes an isometry from $(\mathbb{C}, \varphi^* g)$ to $(\mathbb{C}, g)$ (or whatever the domains of $\mathbf{w}$ and $\mathbf{z}$ may be).

The discussion so far applies to a diffeomorphism in any dimension, but now we use in addition that we are dealing with a biholomorphic map between domains in $\mathbb{C}$. Then the differential $J_{\varphi,\mathbf{w}}$ is simply multiplication by the non-zero complex number $\mathrm{d}\mathbf{z}/\mathrm{d}\mathbf{w}\,(\mathbf{w})$. It follows that

$$(\varphi^* |\mathrm{d}\mathbf{z}|^2)_{\mathbf{w}} = \left|\frac{\mathrm{d}\mathbf{z}}{\mathrm{d}\mathbf{w}}\right|^2 (\mathbf{w}) \cdot |\mathrm{d}\mathbf{w}|^2.$$

In particular, the pull-back $\varphi^*\langle\,.\,,.\,\rangle$ of the euclidean metric is conformally equivalent to that metric. In the proof of Theorem 9.13 we have dropped the point $\mathbf{w}$ from the notation and also – abusing notation in a way that is quite customary – the pull-back map φ^*.

Example 9.15 For $\mathbf{z} = \mathbf{w}^2$ we obtain $U(\mathbf{w}) = 4|\mathbf{w}|^2$, which is the Hooke potential for the equation $\mathbf{w}'' = -8\mathbf{w}$, and $V(\mathbf{z}) = -1/4|\mathbf{z}|$, the Newton potential for the equation $\ddot{\mathbf{z}} = -\mathbf{z}/4|\mathbf{z}|^3$.

In order to see the duality between the equations (H) and (K) from Section 8.7, where the force laws involve scaling factors λ, μ, we should start from the identity

$$\left(\frac{\mu}{4} - \frac{\lambda}{8}\left|\frac{\mathrm{d}\mathbf{z}}{\mathrm{d}\mathbf{w}}\right|^2\right)|\mathrm{d}\mathbf{w}|^2 = \frac{1}{16}\left(-2\lambda + 4\mu\left|\frac{\mathrm{d}\mathbf{w}}{\mathrm{d}\mathbf{z}}\right|^2\right)|\mathrm{d}\mathbf{z}|^2.$$

With $\mathbf{z} = \mathbf{w}^2$ we then see that solutions of (H) of energy $h_{\mathrm{H}} = \mu/4$ transform to solutions of (K) of energy $h_{\mathrm{K}} = -2\lambda$, as in Proposition 8.60.

9.2 The Hamilton equations

Our aim now is to transform the Euler–Lagrange equation (EL), which is a system of n second-order equations, into a system of $2n$ first-order equations. Loosely speaking, this is done by introducing $\partial L/\partial v_i$, $i = 1, \ldots, n$, as new variables. For an *ordinary* differential equation of order two such as (CFP) on page 1 one can indeed simply introduce a new variable and a new equation setting it equal to the first derivative, which led to (CFP′). For a system of *partial* differential equations such as (EL) the transformation is a little more subtle.

Introduce additional variables $\mathbf{p} = (p_1, \ldots, p_n) \in \mathbb{R}^n$ and consider the system of equations

$$p_i - \frac{\partial L}{\partial v_i}(t, \mathbf{q}, \mathbf{v}) = 0, \quad i = 1, \ldots, n, \tag{9.6}$$

for $(t, \mathbf{q}, \mathbf{v}, \mathbf{p}) \in [a, b] \times \Omega \times \mathbb{R}^n \times \mathbb{R}^n$. By the implicit function theorem, this system can locally be solved for $v_1, \ldots, v_n$, i.e.

$$v_i = v_i(t, \mathbf{q}, \mathbf{p}), \quad i = 1, \ldots, n,$$

near any solution of (9.6), provided the **Legendre condition**

$$\det\left(\frac{\partial^2 L}{\partial v_i\, \partial v_j}\right) \neq 0$$

is satisfied at that point.

Geometrically speaking, the solution set of (9.6) is a graph over the set $[a, b] \times \Omega \times \mathbb{R}^n$, so it is a $(2n + 1)$-dimensional submanifold of

$$[a, b] \times \Omega \times \mathbb{R}^n \times \mathbb{R}^n \subset \mathbb{R}^{3n+1}.$$

At points of this graph where the Legendre condition is satisfied, this submanifold can be parametrised in terms of the variables $(t, \mathbf{q}, \mathbf{p})$.

In these local charts, we may regard $L(t, \mathbf{q}, \mathbf{v})$ as a function of $(t, \mathbf{q}, \mathbf{p})$ by expressing $\mathbf{v}$ in terms of these coordinates. However, the derivatives of L will then involve partial derivatives of $\mathbf{v}$ with respect to $(t, \mathbf{q}, \mathbf{p})$, and not much has been gained. By modifying L, we can ensure that these derivatives become irrelevant.

Notation From now on we shall use a convenient shorthand for expressions that have a formal similarity with the standard scalar product. For example, we read $\mathbf{pq}$ as $\sum_i p_i q_i$ and $\mathbf{p}\, \mathrm{d}\mathbf{q}$ as $\sum_i p_i\, \mathrm{d}q_i$. The same convention will apply to derivatives. For instance, $(\partial L/\partial \mathbf{q})\, \mathrm{d}\mathbf{q}$ is to be read as $\sum_i (\partial L/\partial q_i)\, \mathrm{d}q_i$.

On $[a,b] \times \Omega \times \mathbb{R}^n \times \mathbb{R}^n$ we define a function H by

$$H(t, \mathbf{q}, \mathbf{p}, \mathbf{v}) = \mathbf{p}\mathbf{v} - L(t, \mathbf{q}, \mathbf{v}).$$

We compute the total differential of this function:

$$\mathrm{d}H = \mathbf{v}\,\mathrm{d}\mathbf{p} + \mathbf{p}\,\mathrm{d}\mathbf{v} - \frac{\partial L}{\partial t}\,\mathrm{d}t - \frac{\partial L}{\partial \mathbf{q}}\,\mathrm{d}\mathbf{q} - \frac{\partial L}{\partial \mathbf{v}}\,\mathrm{d}\mathbf{v}.$$

Along the submanifold defined by (9.6), this simplifies to

$$\mathrm{d}H = -\frac{\partial L}{\partial t}\,\mathrm{d}t - \frac{\partial L}{\partial \mathbf{q}}\,\mathrm{d}\mathbf{q} + \mathbf{v}\,\mathrm{d}\mathbf{p},$$

so, in local charts for this submanifold where $\mathbf{v} = \mathbf{v}(t, \mathbf{q}, \mathbf{p})$, the derivatives of the **Hamiltonian function**

$$H(t, \mathbf{q}, \mathbf{p}) := \mathbf{p}\mathbf{v}(t, \mathbf{q}, \mathbf{p}) - L(t, \mathbf{q}, \mathbf{v}(t, \mathbf{q}, \mathbf{p}))$$

become much simpler than those of L. Comparison with the total differential

$$\mathrm{d}H = \frac{\partial H}{\partial t}\,\mathrm{d}t + \frac{\partial H}{\partial \mathbf{q}}\,\mathrm{d}\mathbf{q} + \frac{\partial H}{\partial \mathbf{p}}\,\mathrm{d}\mathbf{p}$$

shows that in such a chart we have

$$\left.\begin{array}{rcl} \mathbf{v} & = & \dfrac{\partial H}{\partial \mathbf{p}}, \\ \dfrac{\partial L}{\partial \mathbf{q}} & = & -\dfrac{\partial H}{\partial \mathbf{q}}. \end{array}\right\} \qquad (9.7)$$

Now let $t \mapsto \mathbf{q}(t)$ be a solution of the Euler–Lagrange equation, that is,

$$\frac{\mathrm{d}}{\mathrm{d}t}\left(\frac{\partial L}{\partial \mathbf{v}}(t, \mathbf{q}, \dot{\mathbf{q}})\right) - \frac{\partial L}{\partial \mathbf{q}}(t, \mathbf{q}, \dot{\mathbf{q}}) = \mathbf{0}.$$

Notice the slight abuse of notation that we write $\mathbf{q}$ both for an independent variable in the expression $\partial L/\partial \mathbf{q}$ and for a function of time (where previously we wrote α). However, in this notation the following equations become much more memorable.

We may then likewise regard $\mathbf{p}$ as a function of time:

$$\mathbf{p}(t) = \frac{\partial L}{\partial \mathbf{v}}(t, \mathbf{q}(t), \dot{\mathbf{q}}(t)).$$

From (9.7) we now deduce that the function $t \mapsto (\mathbf{q}(t), \mathbf{p}(t))$ satisfies the **Hamilton equations**

$$\left.\begin{array}{rcl} \dot{\mathbf{q}} & = & \dfrac{\partial H}{\partial \mathbf{p}}(t, \mathbf{q}, \mathbf{p}) \\ \dot{\mathbf{p}} & = & -\dfrac{\partial H}{\partial \mathbf{q}}(t, \mathbf{q}, \mathbf{p}), \end{array}\right\}$$

which are equations of precisely the type discussed in the introduction to this chapter. Such a system of equations is also called a **Hamiltonian system**.

This transformation process from L to H (known as a **Legendre transformation**) can be reversed. Locally near points where

$$\det\left(\frac{\partial^2 H}{\partial p_i \partial p_j}\right) \neq 0,$$

a solution of the Hamilton equations can be transformed into a solution of the Euler–Lagrange equation.

Example 9.16 For an autonomous Lagrangian function $L = L(\mathbf{q}, \mathbf{v})$, the Hamiltonian function $H(\mathbf{q}, \mathbf{p}) = \mathbf{p}\mathbf{v} - L(\mathbf{q}, \mathbf{v})$ is precisely the constant of motion found in equation (EL′) of Lemma 9.9.

For the Lagrangian function in Example 9.5, we have $\mathbf{p} = \partial L/\partial \mathbf{v} = m\mathbf{v}$, thus

$$H(\mathbf{q}, \mathbf{p}) = \mathbf{p}\mathbf{v} - L(\mathbf{q}, \mathbf{v}) = \frac{\mathbf{p}^2}{m} - \frac{\mathbf{p}^2}{2m} + V(\mathbf{q}) = \frac{\mathbf{p}^2}{2m} + V(\mathbf{q}),$$

the total energy of the system.

If we set $E(t) = H(t, \mathbf{q}(t), \mathbf{p}(t))$, then the Hamilton equations yield

$$\dot{E} = \frac{\partial H}{\partial t} + \frac{\partial H}{\partial \mathbf{q}}\dot{\mathbf{q}} + \frac{\partial H}{\partial \mathbf{p}}\dot{\mathbf{p}} = \frac{\partial H}{\partial t}.$$

A Hamiltonian system described by a Hamiltonian function $H = H(\mathbf{q}, \mathbf{p})$ that does not depend explicitly on time is called **autonomous**. Our computation shows that, in this case, H is a constant of motion.

The range of the variable $\mathbf{q}$ is called the **configuration space** of the Hamiltonian system. For instance, for the Hamiltonian formulation of (PCR3B), where (q_1, q_2) coincide with the rotating coordinates (x, y), the configuration space is $\mathbb{R}^2 \setminus \{(-\mu, 0), (1 - \mu, 0)\}$.

In the classical situation described in Example 9.5, the p_i equal the momenta with respect to the coordinates q_i, that is, $p_i = mv_i$. In a general Hamiltonian system, the p_i are referred to as the **generalised momenta**, and the range of positions and momenta $(\mathbf{q}, \mathbf{p})$ is called the **phase space**.

Summarising, we have the following result.

Proposition 9.17 *In an autonomous Hamiltonian system, the Hamiltonian function H is a constant of motion, i.e. every solution curve in phase space is contained in a level set of H.* □

9.3 Canonical transformations

Write points $\mathbf{z} \in \mathbb{R}^{2n}$ in the form

$$\mathbf{z} = \begin{pmatrix} \mathbf{q} \\ \mathbf{p} \end{pmatrix}, \quad \mathbf{q} = \begin{pmatrix} q_1 \\ \vdots \\ q_n \end{pmatrix}, \quad \mathbf{p} = \begin{pmatrix} p_1 \\ \vdots \\ p_n \end{pmatrix}.$$

Let $H\colon \mathbb{R}^{2n} \to \mathbb{R}$ be a differentiable function. By ∇H we denote the gradient of H, regarded as a column vector, i.e.

$$\nabla H = \left(\frac{\partial H}{\partial \mathbf{z}} \right)^{\mathrm{t}} = \begin{pmatrix} \partial H / \partial q_1 \\ \vdots \\ \partial H / \partial p_n \end{pmatrix}.$$

Let $E = E_n$ be the $n \times n$ unit matrix, and J the $2n \times 2n$ matrix

$$J = \begin{pmatrix} 0 & -E \\ E & 0 \end{pmatrix}.$$

Observe that $J^2 = -E_{2n}$. Under the identification of $\mathbb{R}^{2n}$ with $\mathbb{R}^n \oplus \mathrm{i}\mathbb{R}^n \equiv \mathbb{C}^n$, application of the matrix J corresponds to multiplication by i.

We can then write the (autonomous) Hamilton equations as

$$\begin{pmatrix} -\dot{\mathbf{p}} \\ \dot{\mathbf{q}} \end{pmatrix} = \nabla H,$$

or, even more succinctly,

$$\boxed{J\dot{\mathbf{z}} = \nabla H.} \tag{9.8}$$

Now consider a diffeomorphism $\mathbf{w} \mapsto \mathbf{z} = \varphi(\mathbf{w})$ of $\mathbb{R}^{2n}$ and the pull-back K of the Hamiltonian function H under this diffeomorphism, i.e.

$$K(\mathbf{w}) := H(\varphi(\mathbf{w})).$$

We compute

$$\begin{aligned} \nabla K(\mathbf{w}) = \left(\frac{\partial K}{\partial \mathbf{w}}(\mathbf{w}) \right)^{\mathrm{t}} &= \left(\frac{\partial H}{\partial \mathbf{z}}(\mathbf{z}) \cdot \frac{\partial \varphi}{\partial \mathbf{w}}(\mathbf{w}) \right)^{\mathrm{t}} \\ = \left(\frac{\partial \varphi}{\partial \mathbf{w}} \right)^{\mathrm{t}} \nabla H &= \left(\frac{\partial \varphi}{\partial \mathbf{w}} \right)^{\mathrm{t}} J\dot{\mathbf{z}} \\ = \left(\frac{\partial \varphi}{\partial \mathbf{w}} \right)^{\mathrm{t}} J \frac{\partial \varphi}{\partial \mathbf{w}} \dot{\mathbf{w}}. & \end{aligned}$$

This motivates the following definition.

Definition 9.18 A diffeomorphism φ of open subsets of $\mathbb{R}^{2n}$ is called a **canonical transformation** if its Jacobian matrix $\Phi := \partial\varphi/\partial\mathbf{w}$ satisfies

$$\boxed{\Phi^{\mathrm{t}} J \Phi = J} \tag{9.9}$$

at every point $\mathbf{w}$ in its domain of definition.

Our computation shows that canonical transformations are precisely those diffeomorphisms that preserve the form of the Hamilton equations. Notice that condition (9.9) taken by itself guarantees that Φ is invertible, so a map φ with such a Jacobian is automatically a local diffeomorphism.

Definition 9.19 A real $2n \times 2n$ matrix Φ satisfying equation (9.9) is called **symplectic**.

Examples 9.20 (1) The matrix J is symplectic, since $J^2 = -E_{2n}$ and $J^{\mathrm{t}} = -J$, hence $J^{\mathrm{t}} J J = J$.

(2) The 2×2 matrix $\left(\begin{smallmatrix} \lambda & 0 \\ 0 & 1/\lambda \end{smallmatrix}\right)$ is symplectic for any $\lambda \in \mathbb{R} \setminus \{0\}$.

Remark 9.21 The symplectic $2n \times 2n$ matrices form a group under matrix multiplication, see Exercise 9.4.

We now want to rewrite the Hamilton equation (9.8) as an equation in differential forms. Write $\partial_{q_i}, \partial_{p_i}$ for the unit tangent vector at a given point in $\mathbb{R}^{2n}$ in the coordinate direction of q_i, p_i, respectively. Recall that on a tangent vector

$$\begin{pmatrix} \mathbf{v} \\ \mathbf{u} \end{pmatrix} = \sum_{i=1}^{n} v_i \, \partial_{q_i} + \sum_{i=1}^{n} u_i \, \partial_{p_i}$$

the differentials $\mathrm{d}q_j$ and $\mathrm{d}p_j$ evaluate as

$$\mathrm{d}q_j \begin{pmatrix} \mathbf{v} \\ \mathbf{u} \end{pmatrix} = v_j, \quad \mathrm{d}p_j \begin{pmatrix} \mathbf{v} \\ \mathbf{u} \end{pmatrix} = u_j.$$

The wedge product of the two 1-forms $\mathrm{d}p_i$ and $\mathrm{d}q_i$ is the 2-form which evaluates on two vector fields $\mathbf{Z}, \mathbf{Z}'$ on $\mathbb{R}^{2n}$ as

$$(\mathrm{d}p_i \wedge \mathrm{d}q_i)(\mathbf{Z}, \mathbf{Z}') = \mathrm{d}p_i(\mathbf{Z})\, \mathrm{d}q_i(\mathbf{Z}') - \mathrm{d}q_i(\mathbf{Z})\, \mathrm{d}p_i(\mathbf{Z}').$$

Definition 9.22 The **canonical symplectic form** on $\mathbb{R}^{2n}$ is the differential 2-form

$$\omega := \sum_{i=1}^{n} \mathrm{d}p_i \wedge \mathrm{d}q_i.$$

With the notational convention introduced on page 176 we may write this canonical symplectic form as $\mathbf{dp} \wedge \mathbf{dq}$.

Lemma 9.23 *The Hamilton equation*

$$J\begin{pmatrix}\dot{\mathbf{q}}\\ \dot{\mathbf{p}}\end{pmatrix} = \nabla H(\mathbf{q},\mathbf{p})$$

is equivalent to

$$\boxed{\begin{pmatrix}\dot{\mathbf{q}}\\ \dot{\mathbf{p}}\end{pmatrix} = \mathbf{X}_H(\mathbf{q},\mathbf{p}),}$$

where the **Hamiltonian vector field** $\mathbf{X}_H$ *is defined by*

$$\boxed{\omega(\mathbf{X}_H, .) = -\mathrm{d}H.}$$

Proof From

$$\mathrm{d}H = \frac{\partial H}{\partial \mathbf{q}}\,\mathrm{d}\mathbf{q} + \frac{\partial H}{\partial \mathbf{p}}\,\mathrm{d}\mathbf{p}$$

and the defining equation for $\mathbf{X}_H$ we find

$$\mathbf{X}_H = \begin{pmatrix} | \\ \partial H/\partial \mathbf{p} \\ | \\ | \\ -\partial H/\partial \mathbf{q} \\ | \end{pmatrix}, \quad \text{and further} \quad J\mathbf{X}_H = \begin{pmatrix} | \\ \partial H/\partial \mathbf{q} \\ | \\ | \\ \partial H/\partial \mathbf{p} \\ | \end{pmatrix} = \nabla H,$$

as we wanted to show. □

As an immediate consequence of this lemma, we obtain a new proof of Proposition 9.17, which we rephrase in the language of the present section.

Proposition 9.24 *The autonomous Hamiltonian function H is a constant of motion of the corresponding Hamiltonian system, i.e. the flow lines of the Hamiltonian vector field $\mathbf{X}_H$ are tangent to the level sets of H.*

Proof This follows from the skew-symmetry of the differential 2-form ω:

$$\mathrm{d}H(\mathbf{X}_H) = -\omega(\mathbf{X}_H, \mathbf{X}_H) = 0.$$ □

Given a diffeomorphism φ between open subsets of $\mathbb{R}^{2n}$ (or in fact any differentiable map), the pull-back map φ^* on differential forms is defined just like for Riemannian metrics in Remark 9.14.

Proposition 9.25 *The transformation $\varphi\colon (\mathbf{Q},\mathbf{P}) \mapsto (\mathbf{q},\mathbf{p})$ is canonical if and only if $\varphi^*(\mathrm{d}\mathbf{p} \wedge \mathrm{d}\mathbf{q}) = \mathrm{d}\mathbf{P} \wedge \mathrm{d}\mathbf{Q}$.*

Proof Let $\mathbf{Z} = \begin{pmatrix}\mathbf{X}\\ \mathbf{Y}\end{pmatrix}$ be a vector field on $\mathbb{R}^{2n}$, where $\mathbf{X}, \mathbf{Y}$ are $\mathbb{R}^n$-valued. We use the corresponding notation for a second vector field $\mathbf{Z}'$ and compute

$$\begin{aligned}(\mathrm{d}\mathbf{p} \wedge \mathrm{d}\mathbf{q})(\mathbf{Z}, \mathbf{Z}') &= \mathbf{YX}' - \mathbf{XY}' \\ &= (\mathbf{X}^{\mathrm{t}}, \mathbf{Y}^{\mathrm{t}})\begin{pmatrix}-\mathbf{Y}'\\ \mathbf{X}'\end{pmatrix} \\ &= \mathbf{Z}^{\mathrm{t}} J \mathbf{Z}'.\end{aligned}$$

With Φ denoting the Jacobian of φ we have

$$\begin{aligned}\varphi^*(\mathrm{d}\mathbf{p} \wedge \mathrm{d}\mathbf{q})(\mathbf{W}, \mathbf{W}') &= (\mathrm{d}\mathbf{p} \wedge \mathrm{d}\mathbf{q})(\Phi\mathbf{W}, \Phi\mathbf{W}') \\ &= (\Phi\mathbf{W})^{\mathrm{t}} J \Phi \mathbf{W}' \\ &= \mathbf{W}^{\mathrm{t}} \Phi^{\mathrm{t}} J \Phi \mathbf{W}'.\end{aligned}$$

So the equality $\varphi^*(\mathrm{d}\mathbf{p} \wedge \mathrm{d}\mathbf{q}) = \mathrm{d}\mathbf{P} \wedge \mathrm{d}\mathbf{Q}$ holds if and only if $\Phi^{\mathrm{t}} J \Phi = J$. □

Remark 9.26 Because of this result, canonical transformations are also referred to as **symplectomorphisms**. In the classical context, 'canonical transformation' is probably the preferred term; the word 'symplectomorphism' is more common in the context of general symplectic manifolds, that is, manifolds with a 2-form which in suitable local coordinates looks like the canonical symplectic form on $\mathbb{R}^{2n}$.

9.4 Equilibrium points and stability

A point $(\mathbf{q}^\circ, \mathbf{p}^\circ) = (q_1^\circ, \ldots, q_n^\circ, p_1^\circ, \ldots, p_n^\circ)$ is called an **equilibrium point** of a given Hamiltonian system

$$\dot{q}_i = \frac{\partial H}{\partial p_i}, \quad \dot{p}_i = -\frac{\partial H}{\partial q_i}, \quad i = 1, \ldots, n,$$

if the partial derivatives of H vanish at that point, in other words, if the constant map $t \mapsto (\mathbf{q}^\circ, \mathbf{p}^\circ)$ is a solution of the Hamilton equations.

Example 9.27 Consider the Hamiltonian description of (PCR3B) given in equations (9.1) and (9.2). The conditions for an equilibrium point are

$$p_1 = -q_2, \quad p_2 = q_1 \quad \text{and} \quad \operatorname{grad}\Phi = \mathbf{0}.$$

Recall that $(q_1, q_2) = (x, y)$ are the rotating coordinates. The vanishing of $\operatorname{grad}\Phi$ characterises the five libration points, see Section 7.3.2. This means that each libration point (q_1°, q_2°) corresponds to a unique equilibrium point

$$(q_1^\circ, q_2^\circ, -q_2^\circ, q_1^\circ),$$

and there are no further equilibrium points.

9.4.1 Lyapunov stability

We measure distances in the phase space of coordinates $(\mathbf{q}, \mathbf{p})$ in terms of the euclidean norm:

$$d((\mathbf{q}, \mathbf{p}), (\mathbf{q}^\circ, \mathbf{p}^\circ)) := \sqrt{\sum_{i=1}^{n}((q_i - q_i^\circ)^2 + (p_i - p_i^\circ)^2)}.$$

Definition 9.28 An equilibrium point $(\mathbf{q}^\circ, \mathbf{p}^\circ)$ is called **Lyapunov stable** if for every $\varepsilon > 0$ there is a $\delta > 0$ such that the following holds: if

$$d((\mathbf{q}(0), \mathbf{p}(0)), (\mathbf{q}^\circ, \mathbf{p}^\circ)) < \delta,$$

then the solution $t \mapsto (\mathbf{q}(t), \mathbf{p}(t))$ is defined for all $t \in \mathbb{R}_0^+$ and

$$d((\mathbf{q}(t), \mathbf{p}(t)), (\mathbf{q}^\circ, \mathbf{p}^\circ)) < \varepsilon \quad \text{for all } t \in \mathbb{R}_0^+.$$

Put into words: given any ε-ball around the equilibrium point, there is a smaller δ-ball around that point such that any trajectory starting in the δ-ball stays inside the ε-ball for all positive times.

Examples 9.29 (1) For the Hamiltonian

$$H(q, p) = \frac{1}{2}(p^2 + q^2),$$

the Hamilton equations are

$$\dot{q} = p, \quad \dot{p} = -q.$$

The solutions are of the form

$$q(t) = b \sin(t - a), \quad p(t) = b \cos(t - a),$$

that is, the solution curves in phase space are circles around the equilibrium point $(0, 0)$. It follows that this point is Lyapunov stable.

(2) For the Hamiltonian

$$H(q, p) = \frac{1}{2}(p^2 - q^2),$$

the Hamilton equations are

$$\dot{q} = p, \quad \dot{p} = q.$$

The solutions are of the form

$$q(t) = p(t) = be^t,$$

or

$$q(t) = -p(t) = b\mathrm{e}^{-t},$$

or

$$q(t) = b\sinh(t-a), \quad p(t) = b\cosh(t-a),$$

or

$$q(t) = b\cosh(t-a), \quad p(t) = b\sinh(t-a),$$

that is, $(0,0)$ is again an equilibrium point, and the solution curves in phase space are hyperbolas and the half-lines along the diagonals $p = \pm q$. All non-constant solutions, except the ones along the diagonal $q = -p$, go to infinity for $t \to \infty$, so $(0,0)$ is not Lyapunov stable.

Both examples describe autonomous systems, so H is a constant of motion by Proposition 9.17. This allows us to recognise $(0,0)$ as a Lyapunov stable equilibrium in example (1) without writing down the solutions explicitly. Likewise, we see that the origin is a Lyapunov stable equilibrium for the Hamiltonian system described by

$$H(\mathbf{q},\mathbf{p}) = \sum_{i=1}^{n}(a_i q_i^2 + b_i p_i^2),$$

provided $a_i, b_i \in \mathbb{R}_0^+$ for $i = 1,\ldots,n$, since in that case the level sets of H are ellipsoids centred at the origin.

9.4.2 Linear stability

Let us try to understand stability in the case of a linear system

$$\dot{\mathbf{h}} = A\mathbf{h},$$

where A is a constant real $n \times n$ matrix, and we are studying solutions $t \mapsto \mathbf{h}(t) \in \mathbb{R}^n$, $t \in \mathbb{R}$. The origin $\mathbf{0}$ is an equilibrium point of this system.

As shown in a first course on ordinary differential equations (see the notes and references section), the components of any solution of this linear equation can be written as linear combinations of

$$t^k\mathrm{e}^{\lambda t}, \quad t^k\mathrm{e}^{at}\cos bt, \quad t^k\mathrm{e}^{at}\sin bt,$$

where λ can be any real eigenvalue, $a \pm ib$ can be any complex eigenvalue of A, and $k \in \mathbb{N}_0$ can take any value from 0 up to $k_{\max} - 1$, defined by the largest Jordan block for the corresponding eigenvalue being a $k_{\max} \times k_{\max}$ matrix. The condition $k_{\max} = 1$ for an eigenvalue is equivalent to saying that the geometric multiplicity of this eigenvalue equals its algebraic multiplicity, i.e. the dimension of the eigenspace (over $\mathbb{C}$) equals the multiplicity of the eigenvalue as a root of the characteristic polynomial $\det(A - \lambda E)$. Notice that the characteristic polynomial is real, so the complex conjugate of any eigenvalue is likewise an eigenvalue. This implies the following result.

Proposition 9.30 *The origin is a stable equilibrium of the linear system* $\dot{\mathbf{h}} = A\mathbf{h}$ *if and only if there are no eigenvalues with positive real part, and* $k_{\max} = 1$ *for all pure imaginary eigenvalues.* □

We now wish to apply this observation to a given equilibrium point $(\mathbf{q}^\circ, \mathbf{p}^\circ)$ of a Hamiltonian system by linearising the equations near that point. First we consider this process of linearisation for a general dynamical system given by a continuous vector field $\mathbf{X}$ defined in a neighbourhood $\Omega \subset \mathbb{R}^n$ of some point $\mathbf{x}^\circ$, with $\mathbf{X}(\mathbf{x}^\circ) = \mathbf{0}$. Then the first-order system $\dot{\mathbf{x}} = \mathbf{X}(\mathbf{x})$ has an equilibrium point at $\mathbf{x}^\circ$. If $\mathbf{X}$ is differentiable at $\mathbf{x}^\circ$, we can linearise it there, i.e. with $\mathbf{x} = \mathbf{x}^\circ + \mathbf{h}$ we can write

$$\mathbf{X}(\mathbf{x}) = A\mathbf{h} + \mathbf{b}(\mathbf{h}),$$

with $A = (A_{ij})$ being the real $n \times n$ matrix with

$$A_{ij} = \frac{\partial X_i}{\partial x_j}(\mathbf{x}^\circ),$$

and $\mathbf{b}$ a remainder term satisfying

$$\lim_{\mathbf{h}\to\mathbf{0}} \frac{\mathbf{b}(\mathbf{h})}{|\mathbf{h}|} = \mathbf{0}.$$

Definition 9.31 An equilibrium point $\mathbf{x}^\circ$ of the dynamical system $\dot{\mathbf{x}} = \mathbf{X}(\mathbf{x})$ is called **infinitesimally stable** if the origin is a Lyapunov stable equilibrium of the linearised system $\dot{\mathbf{h}} = A\mathbf{h}$.

Ideally, one might hope to deduce the Lyapunov stability of an equilibrium point from its infinitesimal stability, since the latter is much easier to verify. In general, however, this implication does not hold, see Exercises 9.9 and 9.10. Here are two known relations between infinitesimal and Lyapunov stability, see (Hale, 1969, III.2) and (Walter, 2000, § 29).

(i) If all the eigenvalues of A have a negative real part, $\mathbf{x}^\circ$ is even **asymptotically stable**, that is, solutions of $\dot{\mathbf{x}} = \mathbf{X}(\mathbf{x})$ with initial point sufficiently close to $\mathbf{x}^\circ$ converge to $\mathbf{x}^\circ$ in forward time.[1]
(ii) If one eigenvalue of A has a positive real part, then $\mathbf{x}^\circ$ is instable.

[1] In other words, if the origin is asymptotically stable for the linearised system $\dot{\mathbf{h}} = A\mathbf{h}$, then $\mathbf{x}^\circ$ is likewise asymptotically stable for the original system $\dot{\mathbf{x}} = \mathbf{X}(\mathbf{x})$.

We now return to studying the stability of an equilibrium point $(\mathbf{q}^\circ, \mathbf{p}^\circ)$ of a Hamiltonian system, where we assume the Hamiltonian function to be of class at least C^2. Define ε_i, δ_i by

$$q_i = q_i^\circ + \varepsilon_i, \quad p_i = p_i^\circ + \delta_i, \quad i = 1, \dots, n.$$

The linear approximation of the Hamilton equations at the equilibrium point is then given by

$$\begin{pmatrix} \dot{\boldsymbol{\varepsilon}} \\ \dot{\boldsymbol{\delta}} \end{pmatrix} = \begin{pmatrix} A & B \\ -C & -D \end{pmatrix} \begin{pmatrix} \boldsymbol{\varepsilon} \\ \boldsymbol{\delta} \end{pmatrix}, \tag{9.10}$$

where

$$A_{ij} = \left(\frac{\partial^2 H}{\partial p_i \, \partial q_j} \right)^\circ, \quad B_{ij} = \left(\frac{\partial^2 H}{\partial p_i \, \partial p_j} \right)^\circ, \quad C_{ij} = \left(\frac{\partial^2 H}{\partial q_i \, \partial q_j} \right)^\circ, \quad D_{ij} = \left(\frac{\partial^2 H}{\partial q_i \, \partial p_j} \right)^\circ;$$

here the superscript $^\circ$ denotes evaluation at the point $(\mathbf{q}^\circ, \mathbf{p}^\circ)$.

The characteristic roots are those values of λ for which the determinant

$$\begin{vmatrix} A - \lambda E & B \\ -C & -D - \lambda E \end{vmatrix}$$

vanishes. Changing the sign of the second row of block entries changes the sign of the determinant by $(-1)^n$, so we may equally study the zeros of the polynomial

$$\chi(\lambda) := \begin{vmatrix} A - \lambda E & B \\ C & D + \lambda E \end{vmatrix}.$$

Lemma 9.32 *The polynomial χ is even, i.e. it only contains multiples of $1, \lambda^2, \lambda^4, \dots, \lambda^{2n}$.*

Proof We need to show that $\chi(-\lambda) = \chi(\lambda)$. Observing that $D = A^{\mathrm{t}}$ and that B, C are symmetric matrices, and using the invariance of the determinant under transposition, we compute

$$\begin{aligned}
\chi(-\lambda) &= \begin{vmatrix} A + \lambda E & B \\ C & A^{\mathrm{t}} - \lambda E \end{vmatrix} \\
&= \begin{vmatrix} A^{\mathrm{t}} + \lambda E & C \\ B & A - \lambda E \end{vmatrix} \\
&= (-1)^n \begin{vmatrix} C & A^{\mathrm{t}} + \lambda E \\ A - \lambda E & B \end{vmatrix} \\
&= \begin{vmatrix} A - \lambda E & B \\ C & A^{\mathrm{t}} + \lambda E \end{vmatrix} \\
&= \chi(\lambda),
\end{aligned}$$

as we wanted to show. □

This implies that the characteristic roots can be written as $\pm\lambda_1, \dots, \pm\lambda_n \in \mathbb{C}$. We arrive at the following result.

Proposition 9.33 *The equilibrium point* $(\mathbf{q}^\circ, \mathbf{p}^\circ)$ *is infinitesimally stable if and only if all* λ_j *are pure imaginary and have* $k_{\text{max}} = 1$. □

So the stability criterion (i) from page 185 is never applicable to a Hamiltonian system, unfortunately. The stability of the five libration points in (PCR3B) will be analysed in Exercise 9.11.

Notes and references

The presentation of Maupertuis's principle in Section 9.1 was inspired by (Levi, 2014), which is highly recommended if you want to learn more about the geometric ideas behind Hamiltonian mechanics. The Jacobi metric is implicit in Section 4.9 of that text. It is well worth taking a look at the original (Jacobi, 1866, Sechste Vorlesung), where Jacobi writes: "Dieses Princip wird fast in allen Lehrbüchern, auch in den besten, in denen von *Poisson*, *Lagrange* und *Laplace*, so dargestellt, dass es nach meiner Ansicht nicht zu verstehen ist." A good survey article about the Jacobi metric, written in a more advanced differential-geometric language, is (Pin, 1975).

The best introduction to the modern viewpoint on symplectic geometry and topology is (McDuff and Salamon, 1998).

(Albers *et al.*, 2012) is an example of a recent paper on (PCR3B) where the Hamiltonian form of the equations is used; see in particular equation (4.2) in that paper.

The notation H for the Hamiltonian function was introduced by Lagrange (1811) – when William Rowan Hamilton was six years old. Lagrange considered the typical situation where H is the total energy of the mechanical system as in Example 9.16; see Section 10.2 for a description of the Kepler problem as a Hamiltonian system of this kind. It turns out that the letter H actually stands for Christiaan Huygens[2] (1629–1695). On page 241 (Section 2-I-14) of (Lagrange, 1811) we read that "Le premier de ces quatres principes, celui de la conservation des forces vives [conservation of energy], a été trouvé par Huyghens." On page 344 (Section 2-V-21) Lagrange then remarks "[...], d'où résulte l'équation $T + V = H$, laquelle exprime la conservation des forces vives du système." We have encountered the equation $T + V = H$ in Proposition 6.3.

[2] This was pointed out to me by Jacques Féjoz.

(My convention in this text is to use the notation h rather than H when we talk about the constant energy of a given system, or a fixed energy level.)

Good texts on systems of linear differential equations (besides those of Hale and Walter mentioned above) are (Givental, 2001) and (Robinson, 1999).

Exercises

9.1 Show that, in the case of Example 9.5, the constant of motion provided by the left-hand side of (EL′) is the total energy of the particle.

9.2 (a) Show that the condition on α we derived in Example 9.10 can be written as

$$\frac{\sin\theta}{y} = \text{const.},$$

where $y = \alpha(t)$ and θ denotes the angle between the vertical and the tangent direction given by the velocity vector $(1, \dot{\alpha}(t))$.

(b) Show by an elementary geometric argument that semicircles orthogonal to $\partial\mathbb{H}$ satisfy this condition. What is the constant on the right-hand side of the equation in this context?

(c) Write the condition on α as a first-order differential equation $\dot{\alpha} = f(\alpha)$. When does this equation satisfy the Lipschitz condition that allows us to invoke the uniqueness statement in the Picard–Lindelöf theorem? Use this to show that the semicircles orthogonal to $\partial\mathbb{H}$ are the only graph-like geodesics in $\mathbb{H}$.

9.3 Treat Exercise 8.36 in the language of Theorem 9.13.

9.4 (a) Show that any symplectic $2n \times 2n$ matrix Φ is invertible by expressing the inverse matrix explicitly in terms of Φ, Φ^{t} and J.

(b) Show that the symplectic $2n\times 2n$ matrices form a group under matrix multiplication. This group is called the **symplectic group** $\mathrm{Sp}(2n)$.

9.5 Write a real $2n \times 2n$ matrix Φ as a block matrix

$$\Phi = \begin{pmatrix} A & B \\ C & D \end{pmatrix},$$

where A, B, C, D are real $n \times n$ matrices.

(a) Show that Φ is symplectic if and only if

$$A^{\mathrm{t}}C = C^{\mathrm{t}}A, \quad B^{\mathrm{t}}D = D^{\mathrm{t}}B, \quad A^{\mathrm{t}}D - C^{\mathrm{t}}B = E_n.$$

Write Φ^{-1} as a block matrix and conclude that the conditions on A, B, C, D can equivalently be formulated as

$$AB^{\mathrm{t}} = BA^{\mathrm{t}}, \quad CD^{\mathrm{t}} = DC^{\mathrm{t}}, \quad AD^{\mathrm{t}} - BC^{\mathrm{t}} = E_n.$$

(b) With a unitary matrix $Z \in \mathrm{U}(n)$, i.e. a complex $n\times n$ matrix satisfying $Z\overline{Z}^{\mathrm{t}} = E_n$, we can associate the real $2n \times 2n$ matrix

$$\Phi = \Phi(Z) = \begin{pmatrix} X & -Y \\ Y & X \end{pmatrix}$$

by writing $Z = X + \mathrm{i}Y$ with real $n \times n$ matrices X, Y. Show that $\Phi(Z)$ is invertible and that the map $Z \mapsto \Phi(Z)$ is an injective group homomorphism from the unitary group $\mathrm{U}(n)$ to the general linear group $\mathrm{GL}(2n)$ of invertible real $2n \times 2n$ matrices. In this way, $\mathrm{U}(n)$ can be regarded as a subgroup of $\mathrm{GL}(2n)$.

(c) Use (a) and (b) to show that

$$\mathrm{Sp}(2n) \cap \mathrm{O}(2n) = \mathrm{U}(n).$$

Here $\mathrm{O}(2n)$ denotes the group of orthogonal $2n \times 2n$ matrices, and $\mathrm{U}(n)$ is regarded as a subgroup of $\mathrm{GL}(2n)$ as in (b).

9.6 Consider the map

$$\varphi\colon (Q_1, Q_2, P_1, P_2) \longmapsto (q_1, q_2, p_1, p_2)$$

given by

$$\begin{aligned} q_1 &= Q_1 \cos Q_2, \\ q_2 &= Q_1 \sin Q_2, \\ p_1 &= P_1 \cos Q_2 - \frac{P_2}{Q_1} \sin Q_2, \\ p_2 &= P_1 \sin Q_2 + \frac{P_2}{Q_1} \cos Q_2. \end{aligned}$$

(a) Find an open subset of $\mathbb{R}^4$, as large as possible, where the map φ defines a diffeomorphism onto its image.

(b) Show that φ is a canonical transformation. For that it is useful to write the Jacobian of φ as a block matrix, as in the foregoing exercise.

9.7 Consider the Hamiltonian function H from (9.3), describing (PCR3B).

(a) Going back to the definition of Φ in Section 7.3, show that this Hamiltonian can be rewritten as

$$H(q_1, q_2, p_1, p_2) = \frac{1}{2}(p_1^2 + p_2^2) - (q_1 p_2 - q_2 p_1) - U(q_1, q_2) - \frac{1}{2}\mu(1-\mu),$$

where

$$U(q_1, q_2) := \frac{1-\mu}{\rho_1} + \frac{\mu}{\rho_2};$$

here ρ_i denotes the distance of the massless particle at the point (q_1, q_2) from the primary m_i, $i = 1, 2$.

(b) Show that, after the transformation φ from the preceding exercise, the new Hamiltonian function $K = H \circ \varphi$ in the variables Q_1, Q_2, P_1, P_2 is given by

$$K(Q_1, Q_2, P_1, P_2) = \frac{1}{2}\left(P_1^2 + \frac{P_2^2}{Q_1^2}\right) - P_2 - U(Q_1 \cos Q_2, Q_1 \sin Q_2) - \frac{1}{2}\mu(1-\mu).$$

Verify that the Hamilton equations transform as expected.

(c) The description in (a) uses rotating coordinates $(x, y) = (q_1, q_2)$. The relation with the non-rotating coordinates (ξ, η) is given by

$$\xi(t) + \mathrm{i}\eta(t) = \mathrm{e}^{\mathrm{i}t}(x(t) + \mathrm{i}y(t)).$$

Trace back the coordinate transformation to these original coordinates, and use this to show the following identities:

$$P_1^2 + \frac{P_2^2}{Q_1^2} = v^2 = \dot{\xi}^2 + \dot{\eta}^2, \quad P_2 = c = \xi\dot{\eta} - \eta\dot{\xi}.$$

What are the meanings of Q_1, Q_2 and P_1?

9.8 Consider the Hamiltonian system given by the function

$$H(q, p) = \frac{p^2}{2} - \cos q,$$

describing a simple pendulum.

(a) Determine the equilibrium points.

(b) Sketch the so-called **phase portrait** of this system, i.e. the flow lines of the Hamiltonian vector field

$$\mathbf{X}_H = \left(\frac{\partial H}{\partial p}, -\frac{\partial H}{\partial q}\right)$$

in the (q, p)-plane. In particular, draw the level set $\{H = 1\}$.

(c) Discuss the Lyapunov stability of the equilibrium points.

9.9 Consider the Hamiltonian function

$$H(q_1, q_2, p_1, p_2) = \frac{1}{2}(q_1^2 + p_1^2) - (q_2^2 + p_2^2) + \frac{1}{2}(p_1^2 p_2 - p_2 q_1^2 - 2q_1 q_2 p_1).$$

(a) Formulate the Hamilton equations and show that the origin is an equilibrium point.

(b) Show that for each $a \in \mathbb{R}$, a solution of the Hamilton equations is given by

$$q_1 = -\sqrt{2}\,\frac{\cos(t-a)}{t-a}, \quad q_2 = \frac{\cos 2(t-a)}{t-a},$$
$$p_1 = \sqrt{2}\,\frac{\sin(t-a)}{t-a}, \quad p_2 = \frac{\sin 2(t-a)}{t-a}$$

for $t \neq a$.

(c) Conclude that the origin is not Lyapunov stable.

(d) Determine explicitly the solutions of the linearised system around the origin, and in this way show that the origin is infinitesimally stable.

This example is due to Cherry (1928).

9.10 Here is a simpler example of an infinitesimally stable but not Lyapunov stable equilibrium point, albeit in a non-Hamiltonian system. Consider the equations

$$\left.\begin{aligned} \dot{x} &= y + x(x^2+y^2) \\ \dot{y} &= -x + y(x^2+y^2). \end{aligned}\right\}$$

The origin $(x, y) = (0, 0)$ is obviously an equilibrium.

(a) Linearise the system at the origin, and show that the equilibrium is infinitesimally stable.

(b) Find a differential equation for $r := \sqrt{x^2+y^2}$ and use this to show that the origin is not Lyapunov stable.

9.11 In this extended exercise we discuss the stability of the libration points in (PCR3B). Recall that the potential Φ describing this problem in rotating coordinates is given by

$$\Phi(x, y) = \frac{1}{2}(x^2+y^2) + \frac{1-\mu}{\rho_1} + \frac{\mu}{\rho_2} + \frac{1}{2}\mu(1-\mu),$$

where ρ_1 denotes the distance from the massless body at (x, y) to the body of mass $1-\mu$ at the point $(-\mu, 0)$, and ρ_2 the distance to the body of mass μ at the point $(1-\mu, 0)$. Set

$$s = \frac{1-\mu}{\rho_1^3} + \frac{\mu}{\rho_2^3}$$

and

$$a = \frac{\partial^2\Phi}{\partial x^2}, \quad b = \frac{\partial^2\Phi}{\partial x\,\partial y}, \quad c = \frac{\partial^2\Phi}{\partial y^2}.$$

(a) Show that

$$a = 1 + 2s - 3y^2\left(\frac{1-\mu}{\rho_1^5} + \frac{\mu}{\rho_2^5}\right),$$

$$b = 3y\left(\frac{(1-\mu)(x+\mu)}{\rho_1^5} + \frac{\mu(x-1+\mu)}{\rho_2^5}\right),$$

$$c = 1 - s + 3y^2\left(\frac{1-\mu}{\rho_1^5} + \frac{\mu}{\rho_2^5}\right).$$

(b) Verify that the values of a, b, c at the libration points $L_1, \ldots, L_5$ are given by

	a	b	c
L_1	$1+2s$	0	$1-s$
L_2	$1+2s$	0	$1-s$
L_3	$1+2s$	0	$1-s$
L_4	$3/4$	$3\sqrt{3}(1-2\mu)/4$	$9/4$
L_5	$3/4$	$-3\sqrt{3}(1-2\mu)/4$	$9/4$

(c) With each of the libration points we associate the two solutions $t_\pm$ of the quadratic equation

$$t^2 + (4 - a - c)t + ac - b^2 = 0.$$

(1) Show that for L_4 and L_5 both of the roots $t_\pm$ are real, negative, and different from each other if and only if $27\mu(1-\mu) < 1$.

(2) Show that for L_1, L_2 and L_3, assuming that $s > 1$, it never happens that both of the roots $t_\pm$ are real and negative.

(d) At the libration points we have $\partial\Phi/\partial x = 0$. Use this to deduce the identity

$$(1-\mu)\left(\rho_1 - \frac{1}{\rho_1^2}\right)\frac{x+\mu}{\rho_1} + \mu\left(\rho_2 - \frac{1}{\rho_2^2}\right)\frac{x-1+\mu}{\rho_2} = 0$$

at those points. Show further that this can be written as

$$\rho_1(s-1) = \mu\left(1 - \frac{1}{\rho_2^3}\right)$$

at L_1 and as

$$\rho_1(1-s) = \mu\left(1 - \frac{1}{\rho_2^3}\right)$$

at L_2, L_3. Conclude that $s > 1$ at all three Euler points. With the preceding part we see that it is never the case that both of $t_\pm$ are negative at those points.

(e) The Hamiltonian function for (PCR3B) is

$$H(q_1, q_2, p_1, p_2) = \frac{1}{2}(p_1 + q_2)^2 + \frac{1}{2}(p_2 - q_1)^2 - \Phi(q_1, q_2).$$

Write Φ_{ij} for the derivative $\partial^2\Phi/\partial q_i\,\partial q_j$. Show that the polynomial χ from Section 9.4.2 is given by

$$\chi(\lambda) = \begin{vmatrix} -\lambda & 1 & 1 & 0 \\ -1 & -\lambda & 0 & 1 \\ 1 - \Phi_{11} & -\Phi_{12} & \lambda & -1 \\ -\Phi_{12} & 1 - \Phi_{22} & 1 & \lambda \end{vmatrix}.$$

Verify that, with $x = q_1$, $y = q_2$ and $t = \lambda^2$, this gives the polynomial from (c).

Recall from Section 9.4 that the equilibrium points of the Hamiltonian system are $(q_1^\circ, q_2^\circ, -q_2^\circ, q_1^\circ)$, where (q_1°, q_2°) is a libration point. By Proposition 9.33, such an equilibrium point can be infinitesimally stable only if the zeros of χ are pure imaginary.

The libration points L_1, L_2, L_3, where χ has a zero with positive real part, are not Lyapunov stable by the instability criterion (ii) from page 185. The libration points L_4, L_5 are infinitesimally stable for $27\mu(1-\mu) < 1$, since this condition guarantees that the linearisation has four distinct pure imaginary eigenvalues. This condition on μ means $\mu < 1/2 - \sqrt{69}/18 \approx 0.03852$. For the Sun-Jupiter system, the actual value of μ is about 0.001. This means that the Trojan asteroids (see Remark 7.2) are in an infinitesimally stable position.

Deep results in Kolmogorov–Arnol'd–Moser (KAM) theory show that the Lagrange points L_4, L_5 are actually Lyapunov stable for values of μ satisfying $27\mu(1-\mu) < 1$, with the exception of three values where certain resonance phenomena occur, see (Arnol'd *et al.*, 2006, Section 6.3.9). In particular, the Trojans are Lyapunov stable.

For more information on the stability of equilibrium points in Hamiltonian systems, with particular emphasis on the three-body problem, and an introduction to KAM theory, see (Meyer *et al.*, 2009). The importance of the concept of stability is underlined by the fact that Siegel and Moser (1971) devote a good third of their book to a study of this aspect of celestial mechanics.

10

The topology of the Kepler problem

> In the ordinary way I can do great damage to a plate of Jersey *Pais de Mai*, which is a sort of bubble-and-squeak made of potatoes, French beans and onions, fried into a cake and served with little pork sausages, but today the gastric juices simply would not flow and I could only wincingly watch the others eating great store of it while I worked out problems in topology with a hot roll.
>
> Kyril Bonfiglioli, *Something Nasty in the Woodshed*

As we have seen in the preceding chapter, every solution curve of an autonomous Hamiltonian system lies inside an energy hypersurface, i.e. a level set of the Hamiltonian function (Propositions 9.17 and 9.24). In order to understand the dynamics of a given system, such as the question of whether there are any periodic solutions, it is helpful and often essential to understand the topology of these level sets

In this chapter I shall describe the so-called geodesic flow on the 2-sphere as a Hamiltonian system. The connection with the Kepler problem is provided by the discussion in Section 8.3; this connection can also be established via the methods of Chapter 9. Another reason why this model example from Riemannian geometry is of special interest in the context of celestial mechanics lies in the fact that it gives rise to energy hypersurfaces with the same topology as certain energy levels in (PCR3B).

Notably, this discussion will lead us to consider the three-dimensional real projective space $\mathbb{RP}^3$, and I shall present various equivalent descriptions of that space. A homeomorphism between $\mathbb{RP}^3$ and the special orthogonal group $SO(3)$ is constructed with the help of Hamilton's quaternions; an introduction to the quaternion algebra is contained in Section 10.4.

10.1 The geodesic flow on the 2-sphere

As discussed in Section 8.3, the geodesics on the 2-sphere S^2 are the great circles, parametrised proportionally to arc length. For simplicity, we shall now always assume unit speed parametrisation. The set of all unit speed geodesics can be interpreted as the collection of flow lines of a dynamical system on the so-called unit tangent bundle of S^2: given an initial condition comprising a starting point $\mathbf{x}_0 \in S^2$ and an initial velocity vector $\mathbf{y}_0$ of length 1 in the tangent space $T_{\mathbf{x}_0}S^2$ (see Definition 8.21 and Exercise 8.9), there is a unique geodesic $t \mapsto \mathbf{x}(t) \in S^2$ with $\mathbf{x}(0) = \mathbf{x}_0$ and $\dot{\mathbf{x}}(0) = \mathbf{y}_0$. The curve

$$t \longmapsto (\mathbf{x}(t), \dot{\mathbf{x}}(t))$$

may be regarded as a curve in the unit tangent bundle, and it constitutes an integral curve of what is called the geodesic flow.

Rather than going through the formal definitions of the terms used in the preceding paragraph, I now want to describe the geodesic flow on S^2 as a Hamiltonian system on $\mathbb{R}^6$. This is possible thanks to the simple description of S^2 as a submanifold of $\mathbb{R}^3$, which will allow us to think of the unit tangent bundle of S^2 as a submanifold of $\mathbb{R}^6$.

Consider $\mathbb{R}^6 = \mathbb{R}^3 \times \mathbb{R}^3$ with coordinates $(\mathbf{x}, \mathbf{y})$ and the Hamiltonian function

$$H(\mathbf{x}, \mathbf{y}) = \frac{1}{2}(|\mathbf{x}|^2|\mathbf{y}|^2 - \langle \mathbf{x}, \mathbf{y}\rangle^2).$$

The corresponding Hamiltonian system is:

$$\left.\begin{aligned} \dot{\mathbf{x}} &= \frac{\partial H}{\partial \mathbf{y}} = -\langle \mathbf{x}, \mathbf{y}\rangle \mathbf{x} + |\mathbf{x}|^2 \mathbf{y} \\ \dot{\mathbf{y}} &= -\frac{\partial H}{\partial \mathbf{x}} = -|\mathbf{y}|^2 \mathbf{x} + \langle \mathbf{x}, \mathbf{y}\rangle \mathbf{y}. \end{aligned}\right\}$$

Lemma 10.1 *The quantities $|\mathbf{x}|$, $|\mathbf{y}|$ and $\langle \mathbf{x}, \mathbf{y}\rangle$ are constants of motion.*

Proof Let $t \mapsto (\mathbf{x}(t), \mathbf{y}(t))$ be a solution of the Hamiltonian system. Then

$$\begin{aligned} \frac{\mathrm{d}}{\mathrm{d}t}|\mathbf{x}|^2 &= 2\langle \mathbf{x}, \dot{\mathbf{x}}\rangle \\ &= 2\langle \mathbf{x}, -\langle \mathbf{x}, \mathbf{y}\rangle \mathbf{x} + |\mathbf{x}|^2 \mathbf{y}\rangle \\ &= 0. \end{aligned}$$

For $|\mathbf{y}|^2$ the computation is similar. For $\langle \mathbf{x}, \mathbf{y} \rangle$ we have

$$\begin{aligned}\frac{\mathrm{d}}{\mathrm{d}t}\langle \mathbf{x}, \mathbf{y} \rangle &= \langle \dot{\mathbf{x}}, \mathbf{y} \rangle + \langle \mathbf{x}, \dot{\mathbf{y}} \rangle \\ &= \langle -\langle \mathbf{x}, \mathbf{y} \rangle \mathbf{x} + |\mathbf{x}|^2 \mathbf{y}, \mathbf{y} \rangle + \langle \mathbf{x}, -|\mathbf{y}|^2 \mathbf{x} + \langle \mathbf{x}, \mathbf{y} \rangle \mathbf{y} \rangle \\ &= 0.\end{aligned}$$

□

In particular, the flow lines of the Hamiltonian system that start at a point of the subset

$$STS^2 := \{(\mathbf{x}, \mathbf{y}) \in \mathbb{R}^6 \colon \ |\mathbf{x}| = 1, \ |\mathbf{y}| = 1, \ \langle \mathbf{x}, \mathbf{y} \rangle = 0\} \tag{10.1}$$

remain in that subset for all times. Observe that STS^2 may be thought of as the set of unit tangent vectors to S^2: we regard $\mathbf{x}$ as a point on $S^2 \subset \mathbb{R}^3$; any vector $\mathbf{y} \in \mathbb{R}^3$ orthogonal to $\mathbf{x}$ can then be interpreted as an element of the tangent space $T_{\mathbf{x}}S^2$; the unit tangent vectors are those with $|\mathbf{y}| = 1$. We call STS^2 the **unit tangent bundle of** S^2.

Lemma 10.2 *The unit cotangent bundle STS^2 of S^2 is a compact three-dimensional submanifold of $\mathbb{R}^6$.*

Proof The defining equations for STS^2 show that this space is a bounded and closed subset of $\mathbb{R}^6$, and hence compact. We have $STS^2 = f^{-1}(\mathbf{0})$ with $f = (f_1, f_2, f_3) \colon \mathbb{R}^6 \to \mathbb{R}^3$ given by

$$f_1(\mathbf{x}, \mathbf{y}) = |\mathbf{x}|^2 - 1, \quad f_2(\mathbf{x}, \mathbf{y}) = |\mathbf{y}|^2 - 1, \quad f_3(\mathbf{x}, \mathbf{y}) = \langle \mathbf{x}, \mathbf{y} \rangle.$$

The Jacobian of f is

$$\frac{\partial(f_1, f_2, f_3)}{\partial(\mathbf{x}, \mathbf{y})} = \begin{pmatrix} 2x_1 & 2x_2 & 2x_3 & 0 & 0 & 0 \\ 0 & 0 & 0 & 2y_1 & 2y_2 & 2y_3 \\ y_1 & y_2 & y_3 & x_1 & x_2 & x_3 \end{pmatrix},$$

which is of rank 3 along $f^{-1}(\mathbf{0})$, since the three rows are different from $(\mathbf{0}, \mathbf{0})$ and pairwise orthogonal. It follows that $STS^2 \subset \mathbb{R}^6$ is a submanifold of dimension $6 - 3 = 3$. □

By these lemmata, the Hamiltonian system restricts to a flow on STS^2 given by the following equations:

$$\left. \begin{array}{rcl} \dot{\mathbf{x}} & = & \mathbf{y} \\ \dot{\mathbf{y}} & = & -\mathbf{x}. \end{array} \right\}$$

Observe that $(\mathbf{y}, -\mathbf{x})$ is an element of the tangent space $T_{(\mathbf{x},\mathbf{y})}(STS^2)$, as it should be, since this vector is orthogonal to $\operatorname{grad} f_i$ for $i = 1, 2, 3$. In fact, we can write down the flow lines explicitly. The integral curve $t \mapsto (\mathbf{x}(t), \mathbf{y}(t))$ with $(\mathbf{x}(0), \mathbf{y}(0)) = (\mathbf{x}_0, \mathbf{y}_0)$, where $|\mathbf{x}_0| = |\mathbf{y}_0| = 1$ and $\langle \mathbf{x}_0, \mathbf{y}_0 \rangle = 0$, is given by

$$\left.\begin{array}{rcl} \mathbf{x}(t) & = & \mathbf{x}_0 \cos t + \mathbf{y}_0 \sin t \\ \mathbf{y}(t) & = & -\mathbf{x}_0 \sin t + \mathbf{y}_0 \cos t. \end{array}\right\}$$

This means that $\mathbf{x}$ is the unit speed parametrisation of a great circle with starting point $\mathbf{x}_0$ and initial velocity $\mathbf{y}_0$. At any time t, the vector $\mathbf{y}(t)$ is the velocity vector of the geodesic at the point $\mathbf{x}(t)$. This justifies calling this flow on STS^2 the **geodesic flow**.

Remark 10.3 I did not explain how to 'guess' the correct Hamiltonian function that would describe the geodesic flow on STS^2. Observe, however, that our Hamiltonian function H can also be written as $H(\mathbf{x}, \mathbf{y}) = |\mathbf{x} \times \mathbf{y}|^2/2$. Moreover, we notice that for any unit speed geodesic $\mathbf{x}$ on S^2 we have $|\mathbf{x} \times \dot{\mathbf{x}}| = 1$. Since, by Propositions 9.17 and 9.24, the Hamiltonian function has to be a constant of motion, our choice of H is not as miraculous as it may have appeared.

10.2 The Kepler problem as a Hamiltonian system

On the set

$$\{(\mathbf{q}, \mathbf{p}) \in \mathbb{R}^2 \times \mathbb{R}^2 : \ \mathbf{q} \neq \mathbf{0}\}$$

we consider the Hamiltonian function

$$H(\mathbf{q}, \mathbf{p}) = \frac{1}{2}|\mathbf{p}|^2 - \frac{1}{|\mathbf{q}|}.$$

The corresponding Hamiltonian equations are

$$\left.\begin{array}{rcccl} \dot{\mathbf{q}} & = & \dfrac{\partial H}{\partial \mathbf{p}} & = & \mathbf{p} \\ \dot{\mathbf{p}} & = & -\dfrac{\partial H}{\partial \mathbf{q}} & = & -\dfrac{\mathbf{q}}{|\mathbf{q}|^3}, \end{array}\right\}$$

hence $\ddot{\mathbf{q}} = -\mathbf{q}/|\mathbf{q}|^3$. So this gives the Hamiltonian description of the Kepler problem in the plane for a unit mass, with $\mathbf{q}$ describing the position and $\mathbf{p}$ the velocity or momentum; the Hamiltonian function is the total energy of the system.

The energy hypersurface $\{H = -1/2\}$ is not compact. However, as seen in Section 8.3, by including the collision orbits one obtains a flow on a compact manifold, which can be interpreted as the geodesic flow on STS^2. The transformation from the Kepler problem to the geodesic flow can be carried out within the Hamiltonian formalism (see the notes and references section).

So the topology of STS^2, which we are going to discuss in the next two sections, will also be of interest in the context of celestial mechanics.

10.3 The group SO(3) as a manifold

Our aim in this section and the following one will be to identify the manifold STS^2 with both the real projective space $\mathbb{RP}^3$ and the group SO(3) of special orthogonal 3×3 matrices. We begin by showing that the matrix group SO(3) actually *is* a manifold. We regard SO(3) as a subset of the nine-dimensional vector space of all real 3×3 matrices.

Proposition 10.4 *The subset* $\mathrm{SO}(3) \subset \mathbb{R}^9$ *is a three-dimensional submanifold.*

Proof The vector space of *symmetric* 3×3 matrices can be identified with $\mathbb{R}^6$ by taking the entries on and above the diagonal as coordinates. We may then consider the map

$$\begin{array}{rccc} f\colon & \mathbb{R}^9 & \longrightarrow & \mathbb{R}^6 \\ & A & \longmapsto & A^{\mathrm{t}}A. \end{array}$$

Then $f^{-1}(E_3) = \mathrm{O}(3)$. Hence, in order to show that $\mathrm{O}(3) \subset \mathbb{R}^9$ is a manifold of dimension $9 - 6 = 3$, we need to verify that f is a differentiable map with a surjective differential $\mathrm{d}_A f$ in all points $A \in f^{-1}(E_3)$.

Thus, take $A \in \mathrm{O}(3) = f^{-1}(E_3)$, and let h be any real 3×3 matrix. We compute

$$f(A + h) = (A + h)^{\mathrm{t}}(A + h) = f(A) + A^{\mathrm{t}}h + h^{\mathrm{t}}A + h^{\mathrm{t}}h.$$

For given A, the map $h \mapsto A^{\mathrm{t}}h + h^{\mathrm{t}}A$ is linear in h, and the term $h^{\mathrm{t}}h$ is $o(\|h\|)$, that is, the limit of $h^{\mathrm{t}}h/\|h\|$ is the zero matrix as h goes to zero. This means that f is differentiable with differential $\mathrm{d}_A f$ given by

$$\mathrm{d}_A f(h) = A^{\mathrm{t}}h + h^{\mathrm{t}}A.$$

Now, given a symmetric 3×3 matrix k, set $h = Ak/2$. Then

$$\mathrm{d}_A f(h) = \frac{1}{2}(A^{\mathrm{t}}Ak + kA^{\mathrm{t}}A) = k.$$

So $\mathrm{d}_A f$ is indeed surjective, and $\mathrm{O}(3) \subset \mathbb{R}^9$ is a three-dimensional submanifold.

The subset $\mathrm{SO}(3) \subset \mathrm{O}(3)$, given by the additional condition $\det A = 1$, is open and closed, and hence a collection of components of O(3).[1] □

[1] We shall see presently that SO(3) is compact and connected by identifying it with STS^2.

Theorem 10.5 *The manifold* SO(3) *is diffeomorphic to* STS^2.

Proof Recall the definition (10.1) of STS^2 as a subset of $\mathbb{R}^6$. The map from $\mathbb{R}^6$ to the space of real 3×3 matrices defined by

$$(\mathbf{x}, \mathbf{y}) \longmapsto \begin{pmatrix} | & | & | \\ \mathbf{x} & \mathbf{y} & \mathbf{x} \times \mathbf{y} \\ | & | & | \end{pmatrix}$$

is differentiable and injective, and it sends STS^2 onto SO(3), since SO(3) is precisely the set of 3×3 matrices whose columns form a positively oriented orthonormal basis for $\mathbb{R}^3$. The inverse map from SO(3) to STS^2 is given by projecting onto the first two columns, and hence is likewise differentiable. □

10.4 The quaternions

We now wish to construct a homeomorphism[2] between SO(3) and the real projective 3-space $\mathbb{RP}^3$. We shall do this in the language of Hamilton's quaternions.

First, though, I want to give a heuristic argument, which could also be converted into a rigorous proof. Recall from Section 8.6.5 that $\mathbb{RP}^3$ was defined as the quotient space of S^3 under the identification of antipodal points. Since any point in the upper hemisphere has an antipodal partner in the lower hemisphere, we may equivalently think of $\mathbb{RP}^3$ as the quotient of the closed lower hemisphere under the identification of antipodal points in the equatorial 2-sphere. Topologically, the closed lower hemisphere is a closed 3-disc

$$D^3 := \{\mathbf{x} \in \mathbb{R}^3 : \ |\mathbf{x}| \leq 1\}.$$

So we may write

$$\mathbb{RP}^3 = D^3/\mathbf{x} \sim -\mathbf{x} \ \text{ for } \ \mathbf{x} \in \partial D^3 = S^2.$$

In order to relate this space to SO(3), we need to understand the geometry of a matrix $A \in \mathrm{SO}(3)$. The characteristic polynomial χ_A of A is real of degree 3. By the intermediate value theorem, this must have a real zero, i.e. A has a real eigenvalue λ_0. Since A is an isometry for the standard scalar product, we must have $\lambda_0 \in \{\pm 1\}$. Over the complex numbers, the real polynomial χ_A can be factorised as

$$\chi_A(\lambda) = (\lambda - \lambda_0)(\lambda - \lambda_1)(\lambda - \lambda_2)$$

[2] The map we describe is actually a diffeomorphism, but to understand this we would first have to give $\mathbb{RP}^3$ the structure of a differentiable manifold, which is not difficult, but is beyond the aims of this introduction.

with $\lambda_1, \lambda_2 \in \{\pm 1\}$ or $\lambda_1, \lambda_2 \in \mathbb{C} \setminus \mathbb{R}$ complex conjugates of each other. From

$$1 = \det A = \lambda_0 \lambda_1 \lambda_2$$

we deduce that at least one of the eigenvalues must be equal to 1. On the orthogonal complement of the corresponding eigenspace, A is an orientation-preserving isometry, and hence a rotation.

In conclusion: any element of SO(3) is a rotation about some axis in $\mathbb{R}^3$. The axis of rotation is determined by a unit vector along that axis; the angle of rotation may be taken in the interval $[-\pi, \pi]$, with the sign of the angle determined by the right-hand rule. If we choose the opposite unit vector to orient the axis of rotation, the sign of the angle will change. Moreover, rotation through an angle π or $-\pi$ amounts to the same thing.

These considerations show that we have a well-defined map

$$\begin{array}{ccl} \mathrm{SO}(3) & \longrightarrow & \left((\text{ball } D^3_\pi \text{ of radius } \pi)/\mathbf{x} \sim -\mathbf{x} \text{ for } \mathbf{x} \in \partial D^3_\pi\right) = \mathbb{R}\mathrm{P}^3 \\ A & \longmapsto & [(\text{unit vector on axis of rotation}) \cdot (\text{angle of rotation})]. \end{array}$$

This map can be shown to be a homeomorphism.

Now, as promised, we use an alternative argument involving the quaternions.

Definition 10.6 The **quaternion algebra**[3] $\mathbb{H}$ is the vector space $\mathbb{R}^4$ with the standard basis

$$\begin{aligned} \mathbf{1} &:= (1, 0, 0, 0) \\ \mathbf{i} &:= (0, 1, 0, 0) \\ \mathbf{j} &:= (0, 0, 1, 0) \\ \mathbf{k} &:= (0, 0, 0, 1) \end{aligned}$$

and a product structure defined by

$$\mathbf{i}^2 = \mathbf{j}^2 = \mathbf{k}^2 = \mathbf{ijk} = -\mathbf{1}, \tag{10.2}$$

where $\mathbf{1}$ acts as the neutral element of multiplication, and multiplication is associative and distributive over $\mathbb{R}$.

For instance, we have $\mathbf{ij} = -\mathbf{ijk}^2 = \mathbf{k}$, and similarly $\mathbf{ji} = -\mathbf{k}$.

So a general quaternion is of the form

$$\mathbf{a} = a_0\mathbf{1} + a_1\mathbf{i} + a_2\mathbf{j} + a_3\mathbf{k}$$

with $a_0, a_1, a_2, a_3 \in \mathbb{R}$. We identify the subspace of $\mathbb{H}$ spanned by $\mathbf{1}$ with $\mathbb{R}$ and

[3] The notation $\mathbb{H}$ is chosen in honour of William Rowan Hamilton, who discovered the quaternions in 1843. There should be little reason to confuse this with hyperbolic space.

usually write a_0 instead of $a_0\mathbf{1}$. Multiplication of two quaternions is then given by

$$\begin{aligned}(a_0 + a_1\mathbf{i} + a_2\mathbf{j} + a_3\mathbf{k}) \cdot (b_0 + b_1\mathbf{i} + b_2\mathbf{j} + b_3\mathbf{k}) = \\ (a_0b_0 - a_1b_1 - a_2b_2 - a_3b_3) \\ + (a_0b_1 + a_1b_0 + a_2b_3 - a_3b_2)\mathbf{i} \\ + (a_0b_2 + a_2b_0 + a_3b_1 - a_1b_3)\mathbf{j} \\ + (a_0b_3 + a_3b_0 + a_1b_2 - a_2b_1)\mathbf{k}.\end{aligned} \tag{10.3}$$

We call $a_0 \in \mathbb{R}$ the **real part** of $\mathbf{a}$, and $a_1\mathbf{i} + a_2\mathbf{j} + a_3\mathbf{k}$ the **imaginary part**. The quaternion $\overline{\mathbf{a}}$ **conjugate to a** is defined by

$$\overline{\mathbf{a}} = a_0 - a_1\mathbf{i} - a_2\mathbf{j} - a_3\mathbf{k}.$$

The **norm** of **a** is

$$|\mathbf{a}| := \sqrt{\mathbf{a}\overline{\mathbf{a}}} = \sqrt{a_0^2 + a_1^2 + a_2^2 + a_3^2}.$$

The word ‘algebra’ here means a real vector space with a multiplication that is distributive over $\mathbb{R}$ (but not necessarily commutative or associative). The complex numbers $\mathbb{C}$, for instance, are a two-dimensional commutative and associative algebra. But $\mathbb{C}$ has a further important property: any element different from $\mathbf{0}$ has a multiplicative inverse. An algebra with this property is called a **division algebra**.

Proposition 10.7 *The quaternions $\mathbb{H}$ form a four-dimensional real associative and non-commutative division algebra.*

Proof Associativity (which we formulated as part of the definition) follows from the rule (10.3) for multiplying two quaternions once the associativity of the multiplication of the basis elements $\mathbf{i}, \mathbf{j}, \mathbf{k}$ is stipulated. Alternatively, instead of (10.2) one can prescribe the product of any two basis quaternions, i.e. $\mathbf{ij} = -\mathbf{ji} = \mathbf{k}$ and the cyclic permutations of this equation, and thence derive associativity.

The property of $\mathbb{H}$ being a division algebra follows from the explicit description of the multiplicative inverse: for $\mathbf{a} \neq \mathbf{0}$ the product $\mathbf{a}\overline{\mathbf{a}} = |\mathbf{a}|^2$ is real and non-zero, hence

$$\mathbf{a}^{-1} = \frac{\overline{\mathbf{a}}}{|\mathbf{a}|^2}.$$ □

The rules

$$\overline{\mathbf{a} \cdot \mathbf{b}} = \overline{\mathbf{b}} \cdot \overline{\mathbf{a}}$$

and

$$|\mathbf{a} \cdot \mathbf{b}| = |\mathbf{a}| \cdot |\mathbf{b}|$$

for all $\mathbf{a}, \mathbf{b} \in \mathbb{H}$ are easily verified.

In the following lemma we identify the pure imaginary quaternions in a canonical fashion with $\mathbb{R}^3$. By $\langle ., . \rangle$ we denote the standard inner product on $\mathbb{R}^3$; by $\times$, the cross product.

Lemma 10.8 *For* $\mathbf{w}, \mathbf{x} \in \mathbb{R}^3 \subset \mathbb{H}$ *we have*

$$\mathbf{wx} - \mathbf{xw} = 2\mathbf{w} \times \mathbf{x}.$$

If $|\mathbf{w}| = 1$*, then*

$$\mathbf{wxw} = \mathbf{x} - 2\langle \mathbf{w}, \mathbf{x} \rangle \mathbf{w}.$$

Proof The first identity follows from a close look at (10.3). The second identity is linear (over $\mathbb{R}$) in $\mathbf{x}$, and, because of the cyclic symmetry of the multiplication rules in $\mathbf{i}, \mathbf{j}, \mathbf{k}$, it suffices to verify the identity for $\mathbf{x} = \mathbf{i}$. We compute

$$\begin{aligned}
\mathbf{wiw} &= (w_1\mathbf{i} + w_2\mathbf{j} + w_3\mathbf{k})\mathbf{i}(w_1\mathbf{i} + w_2\mathbf{j} + w_3\mathbf{k}) \\
&= (w_1\mathbf{i} + w_2\mathbf{j} + w_3\mathbf{k})(-w_1 + w_2\mathbf{k} - w_3\mathbf{j}) \\
&= (-w_1^2 + w_2^2 + w_3^2)\mathbf{i} - 2w_1w_2\mathbf{j} - 2w_1w_3\mathbf{k} \\
&= (1 - 2w_1^2)\mathbf{i} - 2w_1w_2\mathbf{j} - 2w_1w_3\mathbf{k} \\
&= \mathbf{i} - 2w_1\mathbf{w} \\
&= \mathbf{i} - 2\langle \mathbf{w}, \mathbf{i} \rangle \mathbf{w}.
\end{aligned}$$

□

Under the identification of $\mathbb{H}$ with $\mathbb{R}^4$, the **unit quaternions** (i.e. quaternions of norm 1) constitute a 3-sphere, and we shall write $S^3 \subset \mathbb{H}$ for the set of these unit quaternions. Notice that S^3 actually constitutes a subgroup of $\mathbb{H}$.

For $\mathbf{u} \in S^3$ define

$$\begin{aligned}
f_{\mathbf{u}}: \quad \mathbb{R}^3 &\longrightarrow \mathbb{R}^3 \\
\mathbf{x} &\longmapsto \mathbf{u}\mathbf{x}\overline{\mathbf{u}}.
\end{aligned}$$

Observe that $\mathbf{x} \mapsto f_{\mathbf{u}}(\mathbf{x})$ is linear in $\mathbf{x}$, and the image is indeed in $\mathbb{R}^3$, since

$$\overline{\mathbf{u} \cdot \mathbf{x} \cdot \overline{\mathbf{u}}} = \mathbf{u} \cdot \overline{\mathbf{x}} \cdot \overline{\mathbf{u}} = \mathbf{u} \cdot (-\mathbf{x}) \cdot \overline{\mathbf{u}} = -\mathbf{u} \cdot \mathbf{x} \cdot \overline{\mathbf{u}},$$

where we have used $\overline{\mathbf{x}} = -\mathbf{x}$ for $\mathbf{x} \in \mathbb{R}^3$, and this property characterises elements of $\mathbb{R}^3 \subset \mathbb{H}$.

Theorem 10.9 *The map* f *given by*

$$\mathbf{u} \longmapsto f_{\mathbf{u}}$$

defines a surjective group homomorphism

$$S^3 \longrightarrow \mathrm{SO}(3)$$

with kernel $\{\pm\mathbf{1}\}$, *and it induces a homeomorphism*

$$\mathbb{RP}^3 \longrightarrow \mathrm{SO}(3).$$

Proof We have observed above that $f_\mathbf{u}$ is an endomorphism of $\mathbb{R}^3$.

(1) $f_\mathbf{u}$ is an isometry: the computation

$$|f_\mathbf{u}(\mathbf{x})| = |\mathbf{u}\mathbf{x}\overline{\mathbf{u}}| = |\mathbf{u}| \cdot |\mathbf{x}| \cdot |\overline{\mathbf{u}}| = |\mathbf{x}|$$

shows that $f_\mathbf{u}$ preserves the norm, and by polarisation it preserves the scalar product. From the explicit definition we see that $f_\mathbf{u}$ depends continuously on $\mathbf{u}$, and, since S^3 is connected, the $f_\mathbf{u}$ must lie in a single component of $\mathrm{O}(3)$. We have $f_{\pm\mathbf{1}} = \mathrm{id}_{\mathbb{R}^n}$, so that component must be $\mathrm{SO}(3)$. In **(3)** below we shall see this more geometrically.

(2) The map $\mathbf{u} \mapsto f_\mathbf{u}$ is a homomorphism:

$$f_{\mathbf{uv}}(\mathbf{x}) = \mathbf{uvx}\overline{\mathbf{uv}} = \mathbf{uvx}\overline{\mathbf{v}}\,\overline{\mathbf{u}} = f_\mathbf{u}(f_\mathbf{v}(\mathbf{x})),$$

i.e. $f_{\mathbf{uv}} = f_\mathbf{u} \circ f_\mathbf{v}$.

(3) For $\mathbf{u} \in S^3 \setminus \{\pm\mathbf{1}\}$ we now derive an explicit description of the endomorphism $f_\mathbf{u}$ from which the other stated properties of the map $\mathbf{u} \mapsto f_\mathbf{u}$ follow easily.

Write $\mathbf{u} = u_0 + u_1\mathbf{i} + u_2\mathbf{j} + u_3\mathbf{k}$ with $u_0^2 + u_1^2 + u_2^2 + u_3^2 = 1$ and $u_0 \neq \pm 1$. There is a unique $\theta \in (0, 2\pi)$ and a $\mathbf{w} \in \mathbb{R}^3$ with $|\mathbf{w}| = 1$ such that

$$\mathbf{u} = \cos\frac{\theta}{2} + \sin\frac{\theta}{2}\,\mathbf{w}.$$

With Lemma 10.8 we compute

$$\begin{aligned} f_\mathbf{u}(\mathbf{x}) &= \left(\cos\frac{\theta}{2} + \sin\frac{\theta}{2}\,\mathbf{w}\right)\mathbf{x}\left(\cos\frac{\theta}{2} - \sin\frac{\theta}{2}\,\mathbf{w}\right) \\ &= \cos^2\frac{\theta}{2}\,\mathbf{x} - \sin^2\frac{\theta}{2}\,\mathbf{wxw} + \sin\frac{\theta}{2}\cos\frac{\theta}{2}\,(\mathbf{wx} - \mathbf{xw}) \\ &= \left(\cos^2\frac{\theta}{2} - \sin^2\frac{\theta}{2}\right)\mathbf{x} + 2\sin^2\frac{\theta}{2}\,\langle\mathbf{w},\mathbf{x}\rangle\,\mathbf{w} + 2\sin\frac{\theta}{2}\cos\frac{\theta}{2}\,\mathbf{w}\times\mathbf{x} \\ &= \cos\theta\,\mathbf{x} + (1 - \cos\theta)\,\langle\mathbf{w},\mathbf{x}\rangle\,\mathbf{w} + \sin\theta\,\mathbf{w}\times\mathbf{x}. \end{aligned}$$

From this we see the following.

(i) $f_\mathbf{u}(\mathbf{w}) = \mathbf{w}$.

(ii) For $\mathbf{y} \in \mathbb{R}^3$ with $\langle \mathbf{w}, \mathbf{y} \rangle = 0$ we have

$$f_{\mathbf{u}}(\mathbf{y}) = \cos\theta\, \mathbf{y} + \sin\theta\, \mathbf{w} \times \mathbf{y}.$$

For $|\mathbf{y}| = 1$, the vectors $\mathbf{w}, \mathbf{y}, \mathbf{w} \times \mathbf{y}$ form a right-handed orthonormal basis for $\mathbb{R}^3$, so these two properties of $f_{\mathbf{u}}$ imply that $f_{\mathbf{u}}$ is the rotation through an angle θ about the oriented axis given by $\mathbf{w}$, which shows once again that $f_{\mathbf{u}} \in \mathrm{SO}(3)$. Moreover, the statements about the kernel and the surjectivity of the homomorphism $f\colon S^3 \to \mathrm{SO}(3)$ are then also obvious.

(4) From $f_{-\mathbf{u}} = f_{\mathbf{u}}$ it follows that the map f descends to a bijective map $h\colon \mathbb{R}\mathrm{P}^3 \to \mathrm{SO}(3)$ defined by $h([\mathbf{u}]) := f_{\mathbf{u}}$, i.e. we have the following commutative diagram:

$$\begin{array}{ccc} S^3 & \xrightarrow{\ f\ } & \mathrm{SO}(3) \\ {\scriptstyle\pi}\big\downarrow & & \big\downarrow{\scriptstyle\mathrm{id}} \\ \mathbb{R}\mathrm{P}^3 & \xrightarrow{\ h\ } & \mathrm{SO}(3). \end{array}$$

(5) Given any open subset $U \subset \mathrm{SO}(3)$, we know that $\pi^{-1}(h^{-1}(U)) = f^{-1}(U)$ is open, since f, as observed in **(1)**, is continuous. By the definition of the quotient topology, this means that $h^{-1}(U)$ is open. Hence, h is continuous.

(6) The map h is a homeomorphism: For this we need to show that the inverse map h^{-1} is continuous, i.e. for $V \subset \mathbb{R}\mathrm{P}^3$ open we need to verify that $(h^{-1})^{-1}(V) = h(V)$ is open in $\mathrm{SO}(3)$. The two crucial ingredients are the compactness of $\mathbb{R}\mathrm{P}^3$ and the Hausdorff property[4] of $\mathrm{SO}(3)$. Indeed, if $V \subset \mathbb{R}\mathrm{P}^3$ is open, then $\mathbb{R}\mathrm{P}^3 \setminus V$ is compact as a closed subset of a compact space. Then $h(\mathbb{R}\mathrm{P}^3 \setminus V) \subset \mathrm{SO}(3)$ is compact as the continuous image of a compact set, and hence closed, since $\mathrm{SO}(3)$ is a Hausdorff space. This means that $h(V) = \mathrm{SO}(3) \setminus h(\mathbb{R}\mathrm{P}^3 \setminus V)$ – where of course we use the fact that h is bijective – is open in $\mathrm{SO}(3)$. □

Here is a brief sketch showing how to see that $h\colon \mathbb{R}\mathrm{P}^3 \to \mathrm{SO}(3)$ is actually a diffeomorphism. The real projective space $\mathbb{R}\mathrm{P}^3$ can be given the structure of a differentiable manifold such that the projection map $\pi\colon S^3 \to \mathbb{R}\mathrm{P}^3$ is a local diffeomorphism. Therefore it suffices to show that f is a local diffeomorphism. By the symmetry properties of f, we need to check this only at the point $\mathbf{1} \in S^3$.

First we determine the tangent space $T_E\mathrm{SO}(3)$ of $\mathrm{SO}(3)$ at $E = E_3$. Let $t \mapsto A_t \in \mathrm{SO}(3)$ be a smooth curve with $A_0 = E$. From $A_t^{\mathsf{t}} A_t = E$ we have

$$\dot{A}_0^{\mathsf{t}} A_0 + A_0^{\mathsf{t}} \dot{A}_0 = \mathbf{0},$$

[4] The submanifold $\mathrm{SO}(3) \subset \mathbb{R}^6$ inherits the Hausdorff property from $\mathbb{R}^6$.

i.e. $\dot{A}_0^{\mathrm{t}} + \dot{A}_0 = \mathbf{0}$. This implies that $T_E\mathrm{SO}(3)$ equals the three-dimensional vector space of skew-symmetric 3×3 matrices.

Now let $t \mapsto \mathbf{u}_t \in S^3$ be a smooth curve with $\mathbf{u}_0 = \mathbf{1}$. The element $f_{\mathbf{u}_t} \in \mathrm{SO}(3)$ is given (with respect to the standard basis $\mathbf{i}, \mathbf{j}, \mathbf{k}$ of $\mathbb{R}^3$) by the matrix

$$U_t = \begin{pmatrix} | & | & | \\ \mathbf{u}_t\mathbf{i}\,\overline{\mathbf{u}}_t & \mathbf{u}_t\mathbf{j}\,\overline{\mathbf{u}}_t & \mathbf{u}_t\mathbf{k}\,\overline{\mathbf{u}}_t \\ | & | & | \end{pmatrix}.$$

The velocity vector of the curve $t \mapsto U_t$ at $t = 0$ is

$$\dot{U}_0 = \begin{pmatrix} | & | & | \\ \dot{\mathbf{u}}_0\mathbf{i} + \mathbf{i}\,\dot{\overline{\mathbf{u}}}_0 & \dot{\mathbf{u}}_0\mathbf{j} + \mathbf{j}\,\dot{\overline{\mathbf{u}}}_0 & \dot{\mathbf{u}}_0\mathbf{k} + \mathbf{k}\,\dot{\overline{\mathbf{u}}}_0 \\ | & | & | \end{pmatrix}.$$

The tangent vector $\dot{\mathbf{u}}_0 \in T_{\mathbf{1}}S^3$ can be written as

$$\dot{\mathbf{u}}_0 = h_1\mathbf{i} + h_2\mathbf{j} + h_3\mathbf{k}.$$

From the identity

$$\dot{\mathbf{u}}_0\mathbf{i} + \mathbf{i}\,\dot{\overline{\mathbf{u}}}_0 = 2h_3\mathbf{j} - 2h_2\mathbf{k}$$

and its cyclic permutations we find

$$\dot{U}_0 = \begin{pmatrix} 0 & -2h_3 & 2h_2 \\ 2h_3 & 0 & -2h_1 \\ -2h_2 & 2h_1 & 0 \end{pmatrix}.$$

In this way, any skew-symmetric matrix can be realised, which means that the differential

$$T_{\mathbf{1}}f\colon\, T_{\mathbf{1}}S^3 \longrightarrow T_E\mathrm{SO}(3)$$

is an isomorphism. Hence, by the inverse function theorem, f is a local diffeomorphism around the point $\mathbf{1} \in S^3$.

Notes and references

For further details about the geodesic flow on S^2 as a Hamiltonian system see (Moser and Zehnder, 2005), in particular Example (e) in Section 1.5. The transformation from the Kepler problem in the plane to the geodesic flow on STS^2 within the Hamiltonian formalism is described in Section 1.6 of that book.

The topology of the energy hypersurfaces of (PCR3B), which again are compact manifolds after regularisation, is discussed in (Albers *et al.*, 2012). Topological aspects of the general three-body problem are investigated in (McCord and Meyer, 2000).

(Warner, 1983) is a good introduction to the topology of the classical matrix groups. For instance, that text gives a general argument for showing that $\mathrm{SO}(n)$ is one of the two components of $\mathrm{O}(n)$.

A wealth of information about the quaternions (and other number systems), both mathematical and historical, can be found in (Ebbinghaus *et al.*, 1991). Notably it contains a marvellous chapter by Hirzebruch on the topological proof that real division algebras exist in dimensions 1, 2, 4 and 8 only (the real and complex numbers, the quaternions, and the octonians, respectively).

All topological notions used in the proof of Theorem 10.9 can be found in (Jänich, 2005) or (McCleary, 2006).

Exercises

10.1 The proof of Proposition 10.4 can easily be adapted to show that $\mathrm{O}(n)$ and $\mathrm{SO}(n)$ are submanifolds of the vector space of real $n \times n$ matrices. What is the dimension of these manifolds?

10.2 The **Lorentz group** $\mathrm{O}(3,1)$, which plays an important role in special relativity, is the group of real 4×4 matrices A satisfying the identity

$$A^{\mathrm{t}} DA = D,$$

where D is the diagonal matrix with entries $(1, 1, 1, -1)$. Show that $\mathrm{O}(3,1)$ is a six-dimensional submanifold of the 16-dimensional vector space of all real 4×4 matrices.

10.3 Regard S^1 as the unit circle in $\mathbb{C}$, and $\mathbb{RP}^1$ as the quotient space of S^1 under the identification $z \sim -z$, with the structure of a differential manifold which makes the quotient map $S^1 \to \mathbb{RP}^1$, $z \mapsto [z]$ a local diffeomorphism. Show that the map

$$\begin{array}{ccc} \mathbb{RP}^1 & \longrightarrow & S^1 \\ [z] & \longmapsto & z^2 \end{array}$$

is a diffeomorphism.

10.4 An **action** of a group G on a set X is a map

$$\begin{array}{ccc} G \times X & \longrightarrow & X \\ (g, x) & \longmapsto & g(x) \end{array}$$

with the following properties.

(i) $e(x) = x$ for the unit element $e \in G$ and all $x \in X$.
(ii) $g(h(x)) = (gh)(x)$ for all $g, h \in G$ and $x \in X$.

The special orthogonal group SO(3) acts on the 2-sphere $S^2 \subset \mathbb{R}^3$ in a natural way by

$$\begin{array}{ccccc} \mathrm{SO}(3) & \times & S^2 & \longrightarrow & S^2 \\ (A & , & \mathbf{x}) & \longmapsto & A\mathbf{x}. \end{array}$$

A unit tangent vector $(\mathbf{x}_0, \mathbf{y}_0) \in STS^2$ can be written as a velocity vector $(\mathbf{x}_0, \mathbf{y}_0) = (\mathbf{x}(0), \dot{\mathbf{x}}(0))$ with $t \mapsto \mathbf{x}(t)$ a unit speed curve in S^2. For $A \in \mathrm{SO}(3)$, the **differential** TA of $A\colon S^2 \to S^2$ is the map from STS^2 to itself which sends any velocity vector (of a unit speed curve) to the velocity vector of the image curve under A, i.e.

$$(\mathbf{x}(0), \dot{\mathbf{x}}(0)) \longmapsto \left(A\mathbf{x}(0), \frac{\mathrm{d}}{\mathrm{d}t}(A\mathbf{x})(0)\right).$$

(a) Show that this map is well defined, i.e. it does not depend on the choice of curve $\mathbf{x}$ realising a given unit tangent vector.
(b) Verify that this differential induces an action of SO(3) on STS^2.
(c) Fix a unit tangent vector $(\mathbf{x}_0, \mathbf{y}_0) \in STS^2$. Show that the map

$$\begin{array}{ccc} \mathrm{SO}(3) & \longrightarrow & STS^2 \\ A & \longmapsto & TA(\mathbf{x}_0, \mathbf{y}_0) \end{array}$$

is a diffeomorphism.

References

Albers, P., Frauenfelder, U., van Koert, O., and Paternain, G. P. 2012. Contact geometry of the restricted three-body problem, *Communications on Pure and Applied Mathematics* **65**, 229–263.

Albouy, A. 1995. Symétrie des configurations centrales de quatre corps, *Comptes Rendus de l'Académie des Sciences, Série I – Mathématique* **320**, 217–220.

Albouy, A. 1996. The symmetric central configurations of four equal masses. In *Hamiltonian Dynamics and Celestial Mechanics*, edited by D. G. Saari and Z. Xia, Contemporary Mathematics 198, American Mathematical Society, Providence, RI.

Albouy, A. 2002. Lectures on the two-body problem. In *Classical and Celestial Mechanics – The Recife Lectures*, edited by H. Cabral and F. Diacu, Princeton University Press.

Anderson, J. W. 2005. *Hyperbolic Geometry*, 2nd edition, Springer-Verlag, London.

Apollonius of Perga, third century BC. *Treatise on Conic Sections*, edited in modern notation with introductions, including an essay on the earlier history of the subject by T. L. Heath, Cambridge University Press (1896); Cambridge Library Collection, Cambridge University Press (2014).

Arnol'd, V. I. 1973. *Ordinary Differential Equations*, translated from the Russian and edited by R. A. Silverman, The MIT Press, Cambridge, MA.

Arnol'd, V. I. 1990. *Huygens and Barrow, Newton and Hooke*, Birkhäuser Verlag, Basel.

Arnol'd, V. I., Kozlov, V. V., and Neishtadt, A. I. 2006. *Mathematical Aspects of Classical and Celestial Mechanics*, Dynamical Systems III, 3rd edition, Encyclopaedia of Mathematical Sciences 3, Springer-Verlag, Berlin.

Arnol'd, V. I., and Vasil'ev, V. A. 1989. Newton's *Principia* read 300 years later, *Notices of the American Mathematical Society* **36**, 1148–1154.

Aubrey, J. 1692. *An Idea of Education of Young Gentlemen*. In *Aubrey on Education*, edited by J. E. Stephens, Routledge & Kegan Paul, London (1972).

Aubrey, J. 1693. *Brief Lives*. In *Aubrey's Brief Lives*, edited from the original manuscripts and with a life of John Aubrey by O. L. Dick, Martin Secker & Warburg, London (1949).

Bär, C. 2010. *Elementary Differential Geometry*, translated from the 2001 German original by P. Meerkamp, Cambridge University Press.

Barrow-Green, J. 1997. *Poincaré and the Three Body Problem*, History of Mathematics 11, American Mathematical Society, Providence, RI; London Mathematical Society.

Belbruno, E. A. 1977. Two-body motion under the inverse square central force and equivalent geodesic flows, *Celestial Mechanics* **15**, 467–476.

Bernhard, Th. 1975. *Korrektur*, Suhrkamp Verlag, Frankfurt am Main.

Bierce, A. 1911. *The Devil's Dictionary*, The Folio Society, London (2003).

Bohlin, K. 1911. Note sur le problème des deux corps et sur une intégration nouvelle dans le problème des trois corps, *Bulletin Astronomique* **28**, 113–119.

Bonfiglioli, K. 1976. *Something Nasty in the Woodshed*, Macmillan, London.

Borrelli, R. L., and Coleman, C. S. 2004. *Differential Equations – A Modeling Perspective*, John Wiley & Sons, New York.

Brecht, B. 1938. *Leben des Galilei*. In *Werke*, Große kommentierte Berliner und Frankfurter Ausgabe, Band 5, Aufbau Verlag, Berlin and Weimar; Suhrkamp Verlag, Frankfurt am Main (1988).

Bröcker, Th. 1992. *Analysis III*, BI-Wissenschaftsverlag, Mannheim.

Bützberger, F. 1913. *Über bizentrische Polygone, Steinersche Kreis- und Kugelreihen, und die Erfindung der Inversion*, B. G. Teubner, Leipzig.

Carathéodory, C. 1933. Über die strengen Lösungen des Dreikörperproblems, *Sitzungsberichte, Bayerische Akademie der Wissenschaften, Mathematisch-Naturwissenschaftliche Klasse*, 257–267.

Chandrasekhar, S. 1995. *Newton's* Principia *for the Common Reader*, Oxford University Press.

Charpentier, É., Ghys, É., and Lesne, A. (editors) 2010. *The Scientific Legacy of Poincaré*, translated from the 2006 French original by J. Bowman, History of Mathematics 36, American Mathematical Society, Providence, RI.

Charton, I. 2013. Die Herleitung des ersten Keplerschen Gesetzes mittels projektiver Geometrie, B.Sc. thesis, Universität zu Köln.

Chenciner, A. 2015. Poincaré and the three-body problem. In *Henri Poincaré, 1912–2012 – Poincaré Seminar 2012*, edited by B. Duplantier and V. Rivasseau, Progress in Mathematical Physics 67, Birkhäuser Verlag, Basel.

Chenciner, A., and Montgomery, R. 2000. A remarkable periodic solution of the three-body problem in the case of equal masses, *Annals of Mathematics (2)* **152**, 881–901.

Cherry, T. M. 1928. On periodic solutions of Hamiltonian systems of differential equations, *Philosophical Transactions of the Royal Society of London, Series A*, **227**, 137–221.

Coolidge, J. L. 1920. The fundamental theorem of celestial mechanics, *Annals of Mathematics (2)* **21**, 224.

Coxeter, H. S. M., and Greitzer, S. L. 1967. *Geometry Revisited*, Random House, New York.

Danby, J. M. A. 1992. *Fundamentals of Celestial Mechanics*, 3rd printing of the 2nd edition, Willmann-Bell, Richmond, VA.

Derbes, D. 2001. Reinventing the wheel: Hodographic solutions to the Kepler problems, *American Journal of Physics* **69**, 481–489.

Doyle, A. C. 1891. *The Red-Headed League*. In *The New Annotated Sherlock Holmes*, edited by L. S. Klinger, volume I, W. W. Norton & Company, New York (2005).

Ebbinghaus, H.-D., Hermes, H., Hirzebruch, F., Koecher, M., Mainzer, K., Neukirch, J., Prestel, A., and Remmert, R. 1991. *Numbers*, Graduate Texts in Mathematics 123 – Readings in Mathematics, Springer-Verlag, New York.

Euler, L. 1770. Considérations sur le problème des trois corps, *Mémoires de l'Académie des Sciences de Berlin* **19**, 194–220.

Euler, L. 1960. *Opera Omnia*, series II (Opera Mechanica et Astronomica), volume 25: *Commentationes astronomicae ad theoriam perturbationum pertinentes, volumen primum*, Orell Füssli, Zürich.

Geiges, H. 2008. *An Introduction to Contact Topology*, Cambridge Studies in Advanced Mathematics 109, Cambridge University Press.

Giordano, C. M., and Plastino, A. R. 1998. Noether's theorem, rotating potentials, and Jacobi's integral of motion, *American Journal of Physics* **66**, 989–995.

Givental, A. 2001. *Linear Algebra and Differential Equations*, Berkeley Mathematics Lecture Notes 11, American Mathematical Society, Providence, RI.

van Haandel, M., and Heckman, G. 2009. Teaching the Kepler laws for freshmen, *The Mathematical Intelligencer* **31**, no. 2, 40–44.

Hale, J. K. 1969. *Ordinary Differential Equations*, John Wiley & Sons, New York.

Hall, R. W., and Josić, K. 2000. Planetary motion and the duality of force laws, *SIAM Review* **42**, 115–124.

Hamilton, W. R. 1847. The hodograph, or a new method of expressing in symbolical language the Newtonian law of attraction, *Proceedings of the Royal Irish Academy* **3**, 344–353.

Hampton, M., and Moeckel, R. 2006. Finiteness of relative equilibria of the four-body problem, *Inventiones Mathematicae* **163**, 289–312.

Heckman, G., and de Laat, T. 2012. On the regularization of the Kepler problem, *The Journal of Symplectic Geometry* **10**, 463–473.

Hu, S., and Santoprete, M. 2014. Regularization of the Kepler problem on the 3-sphere, *Canadian Journal of Mathematics* **66**, 760–782.

Jacobi, C. G. J. 1836. Sur le mouvement d'un point et sur un cas particulier du problème des trois corps, *Comptes Rendus de l'Académie des Sciences de Paris* **3**, 59–61.

Jacobi, C. G. J. 1866. *Vorlesungen über Dynamik*, Verlag Georg Reimer, Berlin.

Jänich, K. 2005. *Topologie*, 8th edition, Springer-Verlag, Berlin.

Kasner, E. 1913. Differential-geometric aspects of dynamics. In *The Princeton Colloquium*, American Mathematical Society Colloquium Publications 3, American Mathematical Society, New York; reprinted 2008.

Knörrer, H. 1996. *Geometrie*, Vieweg Verlag, Braunschweig.

Koyré, A. 1952. An unpublished letter of Robert Hooke to Isaac Newton, *Isis* **43**, 312–337.

Lagrange, J.-L. 1772. Essai sur le Problème des trois Corps. Prix de l'Académie des Sciences de Paris. In *Œuvres de Lagrange*, tome VI, Gauthier-Villars, Paris (1873).

Lagrange, J.-L. 1785. Recherches sur la Théorie des perturbations que les comètes peuvent éprouver par l'action des planètes. In *Œuvres de Lagrange*, tome VI, Gauthier-Villars, Paris (1873).

Lagrange, J.-L. 1811. *Mécanique Analytique*, Nouvelle Édition, tome premier, Mme Veuve Courcier, Imprimeur-Libraire pour les Mathématiques, Paris; Cambridge Library Collection, Cambridge University Press (2009).

Laplace, P.-S. 1799. *Traité de Mécanique Céleste*. In *Œuvres de Laplace*, tome IV, Imprimerie Royale, Paris (1845).

Leibniz, G. W. 1710. *Essais de Théodicée sur la bonté de Dieu, la liberté de l'homme et l'origine du mal*, Isaac Troyel, Amsterdam.

Levi, M. 2014. *Classical Mechanics with Calculus of Variations and Optimal Control – An Intuitive Introduction*, Student Mathematical Library 69, American Mathematical Society, Providence, RI.

Levi-Civita, T. 1920. Sur la régularisation du problème des trois corps, *Acta Mathematica* **42**, 99–144.

Levi-Civita, T. 1924. *Fragen der klassischen und relativistischen Mechanik*, Springer-Verlag, Berlin.

Lichtenberg, G. Ch. 1766. Von dem Nutzen, den die Mathematik einem Bel-Esprit bringen kann, *Hannoverisches Magazin*. In *Schriften und Briefe III*, Carl Hanser Verlag, München (1972).

Linton, C. M. 2004. *From Eudoxos to Einstein – A History of Mathematical Astronomy*, Cambridge University Press.

Martin, G. E. 1975. *The Foundations of Geometry and the Non-Euclidean Plane*, Springer-Verlag, New York.

McCleary, J. 2006. *A First Course in Topology – Continuity and Dimension*, Student Mathematical Library 31, American Mathematical Society, Providence, RI.

McCord, C., and Meyer, K. R. 2000. Cross sections in the three-body problem, *Journal of Dynamics and Differential Equations* **12**, 247–271.

McDuff, D., and Salamon, D. 1998. *Introduction to Symplectic Topology*, 2nd edition, Oxford Mathematical Monographs, Oxford University Press.

Meyer, K. R., Hall, G. R., and Offin, D. 2009. *Introduction to Hamiltonian Dynamical Systems and the N-Body Problem*, Applied Mathematical Sciences 90, 2nd edition, Springer-Verlag, New York.

Millman, R. S., and Parker, G. D. 1977. *Elements of Differential Geometry*, Prentice-Hall, Englewood Cliffs, NJ.

Milnor, J. 1983. On the geometry of the Kepler problem, *The American Mathematical Monthly* **90**, 353–365.

Möbius, A. F. 1843. *Die Elemente der Mechanik des Himmels*, Weidmann'sche Buchhandlung, Leipzig. In *Gesammelte Werke*, Vierter Band, Verlag von S. Hirzel, Leipzig (1887).

Moeckel, R. 2014. Central configurations, *Scholarpedia* **9**, no. 4, 10667.

Moeckel, R. 2015. Lectures on central configurations. In *Central Configurations, Periodic Orbits, and Hamiltonian Systems*, Advanced Courses in Mathematics – CRM Barcelona, Birkhäuser Verlag, Basel.

Montgomery, R. 2015. The three-body problem and the shape sphere, *The American Mathematical Monthly* **122**, 299–321.

Moser, J. 1970. Regularization of Kepler's problem and the averaging method on a manifold, *Communications on Pure and Applied Mathematics* **23**, 609–636.

Moser, J., and Zehnder, E. J. 2005. *Notes on Dynamical Systems*, Courant Lecture Notes in Mathematics 12, Courant Institute of Mathematical Sciences, New York; American Mathematical Society, Providence, RI.

Musil, R. 1930/31. *Der Mann ohne Eigenschaften*, neu durchgesehene und verbesserte Auflage, Rowohlt Verlag, Reinbek bei Hamburg (1978).

Needham, T. 1993. Newton and the transmutation of force, *The American Mathematical Monthly* **100**, 119–137.

Newton, I. 1687. *Philosophiae Naturalis Principia Mathematica*, translation by I. B. Cohen and A. Whitman, assisted by J. Buzenz, based on the 3rd edition (1726), The Folio Society, London (2008).

Ortega Ríos, R., and Ureña Alcázar, A. J. 2010. *Introducción a la Mecánica Celeste*, Editorial Universidad de Granada.

Osipov, Yu. S. 1977. The Kepler problem and geodesic flows in spaces of constant curvature, *Celestial Mechanics* **16**, 191–208.

Ostermann, A., and Wanner, G. 2012. *Geometry by Its History*, Undergraduate Texts in Mathematics – Readings in Mathematics, Springer-Verlag, Berlin.

Pin, O. C. 1975. Curvature and mechanics, *Advances in Mathematics* **15**, 269–311.

Pollard, H. 1966. *Mathematical Introduction to Celestial Mechanics*, Prentice-Hall, Englewood Cliffs, NJ.

Robinson, C. 1999. *Dynamical Systems – Stability, Symbolic Dynamics, and Chaos*, Studies in Advanced Mathematics, CRC Press, Boca Raton, FL.

Saari, D. G. 1990. A visit to the Newtonian N-body problem via elementary complex variables, *The American Mathematical Monthly* **97**, 105–119.

Saari, D. G. 2005. *Collisions, Rings, and Other Newtonian N-Body Problems*, CBMS Regional Conference Series in Mathematics 104, American Mathematical Society, Providence, RI.

Secord, J. A. 2014. *Visions of Science – Books and Readers at the Dawn of the Victorian Age*, Oxford University Press.

Siegel, C. L., and Moser, J. K. 1971. *Lectures on Celestial Mechanics*, reprint of the 1971 translation from the German by C. I. Kalme, Classics in Mathematics, Springer-Verlag, Berlin (1995).

Smale, S. 1998. Mathematical problems for the next century, *The Mathematical Intelligencer* **20**, no. 2, 7–15.

Somerville, M. 1831. *Mechanism of the Heavens*, Cambridge Library Collection, Cambridge University Press (2009).

Sommerfeld, A. 1942. *Vorlesungen über Theoretische Physik, Band I: Mechanik*, reprint of the 8th edition (1977), Verlag Harri Deutsch, Frankfurt am Main.

Sundman, K. F. 1912. Mémoire sur le problème des trois corps, *Acta Mathematica* **36**, 105–179.

Szebehely, V. 1967. *Theory of Orbits – The Restricted Problem of Three Bodies*, Academic Press, New York.

Voltaire 1759. *Candide, ou l'optimisme*, J. Cramer, Genève.

Walter, W. 2000. *Gewöhnliche Differentialgleichungen*, 7th edition, Springer-Verlag, Berlin.

Warner, F. W. 1983. *Foundations of Differentiable Manifolds and Lie Groups*, Graduate Texts in Mathematics 94, Springer-Verlag, New York.

Wintner, A. 1941. *The Analytical Foundations of Celestial Mechanics*, Princeton Mathematical Series 5, Princeton University Press.

Index

Printed in the United States
By Bookmasters